Field Guide to the Sedges of the Pacific Northwest

Second Edition

The authors and publisher gratefully acknowledge the generous support of the following organizations, which made possible the writing, photography, and publication of the first edition of the *Field Guide to the Sedges of the Pacific Northwest.*

 USDI Bureau of Land Management

 USDA Forest Service

 U.S. Environmental Protection Agency

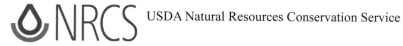 USDA Natural Resources Conservation Service

 Oregon State University Herbarium
Department of Botany and Plant Pathology

 Oregon Flora Project

 Native Plant Society of Oregon

Field Guide to the Sedges of the Pacific Northwest
Second Edition

by
Barbara L. Wilson
Richard E. Brainerd
Danna Lytjen
Bruce Newhouse
& Nick Otting

of the *Carex* Working Group

Oregon State University Press
Corvallis

www.carexworkinggroup.com
the place to go for updates to the identification key
and for other news of Pacific Northwest sedges

Library of Congress Cataloging-in-Publication Data
Field guide to the sedges of the Pacific Northwest / by Barbara L. Wilson ... [et
al.].
   p. cm.
 Includes bibliographical references and index.
 ISBN-13: 978-0-87071-728-4 (alk. paper), ISBN-13: 978-0-87071-729-1
 1. Carex--Oregon--Identification. 2. Carex--Washington (State)--Identification.
I. Wilson, Barbara L., 1951-
 QK495.C997F54 2008
 584'.8409795--dc22

                         2007039207

Oregon State University Press
121 The Valley Library
Corvallis OR 97331-4501
541-737-3166 • fax 541-737-3170
www.osupress.oregonstate.edu

Field Guide to Sedges of the Pacific Northwest

is dedicated to

Danna Lytjen
(1947 - 2006)

Danna was a charter member of the *Carex* Working Group and
was instrumental in bringing this book into existence. She was
an insightful observer of nature, an accomplished field biologist,
a mother, and a grandmother. Her abilities as a scientist, her love
of nature, and her kindness are greatly missed, but they live on in
the hearts of those who knew her.

"I would like to know the grasses and sedges—and care. Then my least journey into the world would be a field trip, a series of happy recognitions."—Annie Dillard

"I would like to go into perfectly new and wild country. I wish to lose myself amid reeds and sedges and wild grasses that have not been touched."—Henry David Thoreau

Botanically, where there's sedge, there's often confusion.—Anon.

Contents

Acknowledgments

The *Carex* Working Group (CWG) began in 1993 when botanists intrigued by taxonomic and ecological issues in the genus *Carex* met at the Oregon State University Herbarium to exchange ideas. At that initial meeting, Dr. Dan Norris suggested that we produce an Atlas of Oregon *Carex*. Peter Zika became the informal leader of the CWG. This careful observer of details inspired us, showed us identification tricks, and led us on field trips. Other early members included Keli Kuykendall, Dr. Francisco (Ankie) Camacho, Jim Oliphant, and Dr. Kate Dwire.

The Native Plant Society of Oregon provided critically important early funding for the *Carex* Working Group's field work and databasing efforts. We are grateful to the society for publishing *The Atlas of Oregon Carex* (Wilson et al. 1999) as its first occasional paper.

Many people and organizations have helped make the Field Guide happen. Major financial support for the project was provided by the Bureau of Land Management. We especially thank Nora Taylor, who first arranged this financial support; Joan Seevers, without whom the project would have died; and Ron Exeter, our patient and effective project manager. BLM Botanists Jean Findley, Claire Hibler, Rick Hall, Ron Halvorson, Lucille Housely, Linda Mazzu, Mark Mousseaux, Bruce Rittenhouse, Tim Rodenkirk, Nancy Sawtelle, Jennifer Wheeler, and Lou Whiteaker made it possible for us to do much of the field work that underlies this book.

Major financial support was also provided by the USDA Forest Service. We thank Russ Holmes for making that possible. Botanists Paula Brooks, David Isle, Jennifer Lippert, Julie Kierstead Nelson, Teresa Ohlsen, Laura Potash Martin, Wayne Rolle, Marty Stein, Maria Ulloa, Gene Yates, and Joan Ziegltrum facilitated field work.

Significant financial support was provided by the U.S. Environmental Protection Agency, with help from Linda Storm. Significant financial support was also provided by the USDA Natural Resources Conservation Service. We thank Kathy Pendergrass for her assistance and Amy Bartow for information about propagation of sedges.

We could not have done this work without the aid of the Oregon State University Herbarium. We thank Dr. Aaron Liston, Director, for early and continued encouragement, and Dr. Richard Halse, Curator, for aid with loans and specimen storage. The Oregon Flora Project (OFP) provided databasing and mapping. The late Dr. Scott Sundberg provided much advice and arranged the initial databasing of *Carex* specimens and sightings. Dr. Linda Hardison and Thea Cook have provided continuing assistance. Katie Mitchell generated the range maps used in the book, managed the databases, assisted in the field, and supervised student workers. We are grateful for her quiet competence.

Our maps are based on databases of herbarium specimens and sightings provided by the Oregon Flora Project (Oregon State University Herbarium), the herbaria of University of Washington and Washington State University, and the other herbaria which contributed data to the Oregon Flora Project.

Plant descriptions are modified from the *Carex* treatment in *Flora of North America* (Ball and Reznicek 2002). Tim Jones provided some of the descriptive material from the database for his on-line interactive key (Jones 2007).

University of Washington Press allowed us to use Jeanne R. Janish's fine drawings from the *Vascular Plants of the Pacific Northwest* (Hitchcock et al. 1969). New York Botanical Garden granted permission to use Harry Charles Creutzburg's drawings from *North American Cariceae* (Mackenzie 1940). Erin Stangel photographed perigynia and herbarium specimens at the Oregon State University Herbarium. Nancy Shaw made available photographs from the *Field Guide to Intermountain Sedges* (Hurd et al. 1998). We thank the following photographers for generously sharing their Carex photos with us: Gerald Carr (Oregon State University Herbarium), David Dister, Susan Farrington, Charles Feddema, Richard Heliwell (USDA Forest Service), Andrew Hipp (Morton Arboretum), Matthias Hoffmann (Martin-Luther-Universität Halle-Wittenberg), Ernst Horak, Emily Kapler, Tom Kaye (Institute for Applied Ecology), Natalie Kirchner, Steve Matson, Keir Morse, Julie Kierstead Nelson (USDA Forest Service), Anton A. Reznicek (University of Michigan Herbarium), Russ Schipper (Audubon Society of Kalamazoo), Forest and Kim Starr (Starr Environmental), Amadej Trnkoczy, Dana Visalli (Methow Biodiversity Project), Leanne Wallis, Fred Weinmann, Louise Wootton (Georgian Court University), Gene Yates (USDA Forest Service), and Peter Zika (University of Washington Herbarium). Stephen Meyers of the Oregon Flora Project formatted the new photos for the 2nd edition.

Dr. Anton A. Reznicek, our sedge guru, read parts of the manuscript and provided advice and insightful discussion. We thank Dr. Robert Naczi (Delaware State University) and Dr. Adolf Ceska (Royal British Columbia Museum) for information about sedges. Dr. Fred Weinmann not only provided information, but led highly productive field trips. Dana Visali and George Wooten provided information about unusual sedges of northern Washington. Dr. Karen Antell (Eastern Oregon University) provided records from northeast Oregon. Peter Lesica provided information about *C. idahoa.* Gay Hunter (Olympic National Park) willingly took last-minute perigynium measurments for us. Dr. Lawrence Janeway (Herbarium at University of California, Chico) was involved in the discovery and description of *C. klamathensis.* Lynda Boyer (Heritage Seedlings, Inc.) provided information about *Carex* seed germination and propagation.

Dr. Mary O'Brien strengthened our early project proposals. Dr. Rhoda Love helped with both early proposals and late editing.

We are grateful to the participants in our sedge identification workshops, whose many suggestions greatly improved the identification key.

We thank Sami Gray for editing and research. We thank Mary Braun and Jo Alexander of OSU Press for their perseverance in assembling the book. Mary Harper of Access Points Indexing prepared the index.

Introduction

Sedges are fun! Repeat this mantra whenever you find yourself facing the challenge of identifying a really difficult sedge. As with many of the more difficult plant groups, a positive attitude and a good sense of humor can smooth the way when you're stuck. Good identification tools help a lot too. Hopefully this book will assist you in becoming more comfortable with identifying sedges.

Carex—sedge—is the largest plant genus in the Pacific Northwest (PNW) with 162 species, and a total of 169 taxa when all subspecies and varieties are counted. These plants play essential roles in a variety of ecosystems and habitats, ranging from the seacoast to alpine slopes, from marshes, lakeshores, and river edges, to dry forests and sagebrush steppe. Particularly in wetlands, sedges often are community dominants, keystone species essential for stabilizing soil, maintaining stream quality, and providing habitat for fish, wildlife and plants. Because of their importance, sedges are frequently used for habitat restoration and enhancement projects. Knowledge of native ranges, habitat requirements, and ecological functions of sedges is essential if these projects are to be successful. Such knowledge begins with identifying the species that are present or that you want to reintroduce.

Sedges have a reputation for being devilishly difficult to identify. Much of the challenge comes from the fact that the plant parts used to identify most sedges are tiny, show great similarity between different species, and have unfamiliar names such as "perigynium." But once you look carefully and master a bit of terminology and plant structure, it turns out that there are some useful "handles" you can use to make sedge identification a manageable, even enjoyable, activity. And having a name for your sedge is a starting point for learning about it and the places where it lives.

This book will be useful for professional botanists, ecologists, wetland scientists, wildlife biologists, and amateur plant enthusiasts. It contains an identification key and descriptions, range maps, photographs, and drawings of all *Carex* taxa known to occur naturally in Oregon and Washington. It also includes information on ecology, and morphology of sedges.

We hope this book will help you to identify most of the sedges that you encounter, and that you won't have to put unidentified "*Carex* sp." on your species list ever again.

How to Use This Book

This field guide provides three main approaches to identifying sedges:

1. Use the identification key (p. 37), checking your result with the pictures and species accounts.

2. If your plant has particularly distinctive features, it may be in the list of sedges with distinctive traits or habitats (p. 412).

3. Leaf through the species accounts to find photos that resemble your sedge. Read the descriptions and use the Identification Tips to compare your specimens to similar species.

If you use methods 2 or 3 to identify your specimen, we strongly recommend that you check the identification by also running the plant through the key.

Consult other regional floras. Although no other single reference includes all the sedges found in Oregon and Washington, each has valuable illustrations, descriptions, and keys. You may find their different perspectives useful. Volume 1 of *Vascular Plants of the Pacific Northwest* (Hitchcock et al. 1969) covers the northern two-thirds of the region covered by this book, and has great drawings. A one-volume summary of this work is *Flora of the Pacific Northwest* (Hitchcock and Cronquist 1973). In southeast Oregon, use the *Intermountain Flora* Volume 6 (Cronquist et al. 1977) and *Field Guide to Intermountain Sedges* (Hurd et al. 1998). In southern Oregon, the *Jepson Manual* sedge treatment (Zika et al. 2012) is useful. *Flora of North America* Volume 23: Cyperaceae (Ball and Reznicek 2002, also available online) provides a continent-wide treatment of sedge species.

An interactive key to *Carex* of North America is available on the Internet (http://www.herbarium.lsu.edu/keys/carex/carex.html). Conspectus tables that compare similar species are available at www.careworkinggroup.com.

Carex Ecology

Sedges perform important ecological functions in a variety of wetland and upland systems. They may be community dominants in an impressive diversity of habitats. They are most valued for stabilizing soil and as food for wild and domesticated animals. A few introduced species may spread as weeds, so understanding their methods of dispersal is important.

Diversity
Carex reach their highest diversity in the north temperate and boreal zones. In the Pacific Northwest (PNW), *Carex* are especially diverse in the alpine zone. Sedges can achieve high species diversity in small areas because they tend to be microhabitat specialists. A few have wide ecological tolerances, and grow in a variety of habitats across the region, but most grow best in a very specific substrate, range of acidity, water and light regime, and elevation. Perhaps most importantly, each species has particular needs for water: deep water all year, deep water in spring and then dry, more or less dry all year, etc. Different species can occupy different zones across the hydrologic gradient around a pond, forming concentric rings from *C. utriculata* in the water through *C. pellita* and *C. scoparia* in seasonally wet marsh, to *C. pachystachya* on the mesic fringes and *C. tumulicola* on the dry slope above.

Climate change will seriously affect the distributions of the many PNW sedges that have limited habitat tolerances. The most vulnerable may be local populations of species such as *C. anthoxanthea, C. circinata, C. pluriflora,* and *C. macrochaeta* that were probably widespread in the PNW at the end of the last ice age, 10,000 years ago. As the climate warmed, these species migrated north, leaving behind the few remnant PNW populations in small, isolated pockets of suitable habitat.

Carex Below Ground
Sedge roots are probably much more interesting than we know. The roots of wetland species are adapted to maintain aerobic metabolism even in water-saturated soils. Air spaces called aerenchyma occur between the cells and allow diffusion of oxygen throughout *Carex* roots, stems, and leaves. This diffusion is efficient enough that excess oxygen may leak from *Carex* roots into adjacent saturated soils, where it oxidizes iron to produce "rust" surrounding the roots (Figure 1).

Most plants depend on an extensive network of mutualistic mycorrhizal fungi to obtain nutrients from the soil and water. However, about half the species in the sedge family lack mycorrhizae and many others are only facultatively mycorrhizal.

Figure 1. Roots of *Carex scoparia* in mud.

In the *Carex* that have mutualistic root fungi, arbuscular mycorrhizae (AM) may be the most common type. Ecotomycorrhizal associations occur in some sedges, such as *Kobresia myosuroides*. Dark septate fungi (DSF) occur in many sedges, especially in habitats where AM do not thrive, such as alpine tundra. The relationship of DSF and their *Carex* hosts needs more study, but appears to benefit both parties. A single plant may host both DSF and AM. Some sedge roots host other fungal species whose role is unclear.

Carex of wetlands frequently lack mycorrhizae because water-logged soil has too little oxygen to promote growth of mycorrhizal fungi. Several sedge adaptations contribute to this independence from mycorrhizal fungi (Muthukumar et al. 2004). Long rhizomes can explore the soil, producing roots in nutrient-rich patches and transporting the nutrients to other parts of the plant. Some sedges obtain nutrients unavailable to mycorrhizal fungi through root clusters in the uppermost soil layer, which contains much litter that could decompose to release nutrients, but which dries out too suddenly and thoroughly to provide good habitat for mycorrhizae. Each season, some sedges transport 80% or more of the nitrogen and phosphorus from their dying leaves to growing shoots. Arctic and alpine sedges need not wait for completion of the slow process of decomposition in cold soils because they can absorb small amino acids such as glycine and use them as a nitrogen source. These sedges also have enzymes on their root surfaces that remove phosphate ions from organic compounds in soil.

Not all below-ground sedge parts are roots. In fact, the most consistently perennial part of a *Carex* is its rhizome, an underground stem that gives rise to both roots and leafy shoots. Rhizomes explore the habitat, transfer nutrients from areas of concentrated resources to the shoots that need them, and determine how sedges compete with neighboring plants.

Sedges, Streams, and Erosion

Sedges exert some of their most important effects on ecosystems by stabilizing soils in prairies, on streambanks, and in wetlands. In wet sedge meadows, 50–75% of the biomass is below ground. Rhizomatous, community-dominant sedges such as *C. nebrascensis, C. aquatilis, C. angustata,* and *C. utriculata* play an important role in stabilizing streamside soils east of the Cascades. Their long, strong rhizomes and roots form a dense network that is tougher than that of dominant riparian zone grasses such as Kentucky Bluegrass (*Poa pratensis*) and Tufted Hairgrass (*Deschampsia cespitosa*). This network prevents both loss of small soil particles and mass wasting of streambanks. Along smaller streams, long rhizomes actually cross below the waterway, knitting the banks together. Even cespitose sedges can hold soil more effectively than one might expect (Figure 2).

Sedges are an important component of the streamside vegetation that provides habitat structure for diverse riparian invertebrates, which are part of the foundation of both aquatic and terrestrial food chains. Sedges also shade stream edges, providing cover for salmon, trout, and other cold-water species. This is especially important in relatively treeless reaches east of the Cascades, such as parts of Oregon's John Day River where *C. nudata* may provide the only shade for fish.

Figure 2. Roots and soil-holding capacity of a small *Carex serratodens.*

Streambank stabilization by sedges can be seriously decreased by grazing livestock or elk. Ungrazed or lightly grazed sedges translocate adequate organic carbon to strengthen and extend rhizomes and roots, but more heavily grazed sedges cannot support the below-ground tissue, and the network of rhizomes and roots deteriorates. In one experiment, ungrazed wet sedge meadows had 60% more below-ground biomass and the soil infiltration rate was more than twice as high as in comparable grazed meadows. Reduction of below-ground biomass can result in stream incision, and the lowered water table further decreases the vigor of the sedge stands, which leads to additional loss of below-ground biomass. Not only does the reduced below-ground biomass lead to stream downcutting, but it also reduces forage production, leading to a spiral of riparian degradation when the animals graze the remaining stubble more intensely. With loss of herbaceous vegetation, grazers may resort to browsing willows with similar effects on their root systems, leading to additional erosion and stream incision. Grazing in sedge meadows while soils are wet can cause soil compaction and churning, directly damaging rhizomes and roots, and reducing their ability to penetrate and stabilize the soil.

Restoring the healthy riparian sedge meadows that stabilize streambanks can be difficult. Along narrow, deeply incised streams, the water table often drops too low to permit seedling establishment.

Too often, projects intended to improve stream habitat include planting of willows in healthy riparian sedge meadows. The planted willows usually die because they are poorly adapted for growth in the deep, anoxic soils. Instead, willows should be planted in disturbed soils to which they are adapted and which would benefit from stabilization by their roots.

Sedges as Forage

Sedges provide nutritious forage for animals capable of digesting cellulose, and can be more nutritious or more easily digested than co-occurring grasses. They are especially important in early spring and late summer because many species green up earlier and remain green later than most grasses. Evergreen species such as *C. geyeri* are important in winter and early spring.

Grazing animals may selectively eat some sedge species, passing by other potential forage plants to do so. For example, cattle select *C. rossii* in sagebrush steppe in spring. Some bison herds consume mainly *C. atherodes* at all seasons of the year. However, many other sedges are tough or have serrate leaf margins that reduce palatability, and are consumed only when preferred forage plants are gone.

The animals that eat *Carex* are diverse. Caterpillars of some butterfly and moth species do so; most are thought to be generalists, eating several species of graminoids, but some are specialists, feeding only on *Carex*. Arctic geese and swans graze so intensively on *C. aquatilis* that they cause shifts in plant communities and in some places even denude the soil. Waterfowl, gallinaceous birds, and small seed-eating birds all eat *Carex* perigynia, sometimes in large quantities. *Carex* foliage is reported to be a significant food source for many mammals: voles, ground squirrels, prairie dogs, beavers, muskrats, pikas, rabbits, bears, bighorn sheep, bison, musk oxen, moose, elk, reindeer, deer, sheep, cattle, and horses. About a third of the diet of bison in Yellowstone National Park is wetland *Carex;* for some Canadian herds, *C. aquatilis* and *C. utriculata* form more than 70% of the winter diet, and for others, *C. atherodes* is the most important food plant at all seasons. Although bears are not specialized to digest cellulose, they may eat large quantities of wetland sedges, especially in spring, digesting mainly the shoot bases and starchy rhizomes. In uplands, bears eat *C. geyeri* in spring.

Effects of herbivory on *Carex* populations depend in part on the growth form of the plants. Cespitose species are dependent on seeds for reproduction. Therefore, although individual plants survive repeated grazing, populations eventually decline even on moderately grazed ranges. For example, *C. petasata,* a cespitose species of sagebrush steppe, declines with grazing and is considered an indicator of range condition, providing 2–3% cover on ranges in excellent condition, less under pressure of grazing. *Carex rossii* and its close relatives can maintain populations despite moderate grazing in part because many of their perigynia are produced on basal spikes too short to be eaten by large animals. Cespitose *C. filifolia* is a source of high-protein forage for livestock and wildlife. It increases under moderate grazing that reduces grass competition, but declines under heavy grazing that prevents seed set or favors monocultures of very short, highly grazing-tolerant grass species. In alpine sites, even very light grazing (one day per year) may cause declines in abundance and cover of *C. filifolia.*

Rhizomatous sedges have the potential to persist and spread in the near-absence of seed production, and therefore withstand grazing better than cespitose species. Some are important forage species (e.g., *C. nebrascensis* and *C. utriculata*), at least after their habitat dries seasonally so that animals have easy access to them. Grazing in fall tends to reduce nutrient storage in rhizomes and may cause harm. Grazing in moist springtime conditions may do less direct harm to plants, but must be balanced against trampling damage to soil and rhizomes. Although the rhizomatous *Carex* may persist under moderately heavy grazing, both above-ground and below-ground biomass are reduced, with potentially harmful effects on their wetland ecosystems.

A few short, wiry, rhizomatous sedges increase on degraded ranges where heavy grazing has reduced competition from grasses. For example, *Carex praegracilis* becomes a community dominant on heavily grazed, seasonally wet, alkaline ranges. In the northern Great Plains, *C. duriuscula* and *C. obtusata* respond the same way in heavily grazed prairies. Dominance by these unusually grazing-tolerant species is accompanied by a reduction of plant diversity, particularly of native species.

Carex and Fire

Effects of fire have been studied in detail in few sedge species. Most *Carex* are well adapted to survive fire because of their growth form and, in many cases, their habitat. *Carex* foliage may burn but the growing points are protected below the soil surface. Survival and cover are greater after low-intensity fires than after high-intensity fires, with the qualification that the intensity that matters is at the rhizome level. Fire that consumes the duff layer can kill upland forest sedges with shallow rhizomes, such as *C. concinnoides*. Many *Carex* have long-lived seed banks that contribute to revegetation after fire and sedges can effectively stabilize soils in burned areas. For example, ten years after the North Waldo Lake fire in the Oregon Cascades, most of the ground cover in uplands is naturally regenerated *C. inops* and *C. rossii* and, in moist drainages, *C. kelloggii.*

Sedges are most susceptible to fire in late spring and early summer, when they are most physiologically active. Fall burns may harm sedges by interrupting storage of nutrients in rhizomes, but some species have responded better to fall burns than spring burns (e.g., *C. duriuscula* in northern prairies). Usually, early spring fires do the least damage. Some species (e.g., *C. rossii*) seem unaffected by season of the fire. Prescribed burning of wet sedge meadows during drought years should be avoided because the dry litter layer and organic soil can burn, killing sedge rhizomes and roots.

Some upland species, such as *C. deflexa, C. geyeri, C. inops,* and *C. rossii,* thrive after fire because of reduced competition and increased soil nutrients. Both survival of mature plants and recruitment from the long-lived soil seed bank contribute to these post-fire sedge populations. Germination of *C. rossii* seeds in the seed bank may be aided by exposure to heat from fires.

Carex geyeri is an upland community dominant in open conifer forest. Prior to settlement, many of its habitats had a history of frequent, low-intensity fires. Its rhizomes survive such fires well and plants may flower the first year after the fire. In the Blue Mountains of Oregon, seed germination also contributes to an overall increase in *C. geyeri* cover after a fire. At some drier sites, competing bunchgrasses may benefit from low-intensity fire more than *C. geyeri,* leading to a reduction in *C. geyeri* cover. Rhizome survival and recruitment from the seed bank both decrease after high-intensity fires, delaying a return to pre-fire cover for seven years or more.

Carex filifolia is an important "bunchgrass" sedge that lives at the dry edge of the ecological niches *Carex* can occupy, in short-grass prairie and sagebrush steppe. It is especially vulnerable if it experiences two or more stressors such as fire, grazing pressure, competition, low precipitation, and poor soil. Severe fires may reduce cover and productivity for fifteen years or more. The effects of low- and moderate-intensity fires vary. If adequate moisture is available, plants recover in two to four years, but low rainfall delays recovery. *Carex filifolia* should not be burned during drought. Recovery after fire depends much more on rhizome survival than recruitment from the seed bank, though both occur. Usually, flowering is reduced the year after a fire, but lower competition and post-fire influx of nutrients may result in unusually high seed production if enough moisture is available. Fire may benefit *C. filifolia* by setting back competing shrubs or grasses, but it may also encourage weed invasion. Resting burned sites from grazing for at least one year facilitates recovery.

Wet sedge meadows may dry out enough to burn in summer or fall, but fire has little effect on deep-rooted wetland species such as *C. aquatilis*. This species may produce more foliage the year after a fire if competing vegetation is killed and additional nutrients are released into the soil. However, prescribed burns have damaged *C. aquatilis* populations already stressed by grazing. If prescribed burns are used as a management tool, pairing them with a rest from grazing will allow root reserves to build up. Ideally, the rest would extend from the season before the burn to two or three years after.

Tussock-forming wetland species also survive fire well. For example, *C. kelloggii* may be one of the few green plants along montane drainages and lakeshores the year after a severe burn. However, wetland sedges that inhabit moss mats or deep organic soils will die if their habitat dries out enough to burn. If global warming leads to drying of *Sphagnum* bogs, increased fire frequency will threaten many wetland sedges and other species.

Seeds and Seed Dispersal

In *Carex,* the perigynium takes over the seed-dispersal function performed by fruits in most flowering plants. All perigynia can float, but some wetland species have special adaptations. These include perigynia with thickened, corky bases (e.g., *C. stipata, C. jonesii*) that cause the perigynium to float with its beak pointing down, the better to stick in mud or vegetation. Others have large, loose, bladder-like perigynia (e.g., *C. utriculata, C. vesicaria*). Small birds, waterfowl, and gallinaceous birds eat *Carex* perigynia, and some of the seeds pass through their digestive systems intact, capable of germinating where they land. Flat, broad perigynia are wind dispersed (e.g., *C. breweri*). This method is common in section *Ovales* (e.g., *C. microptera*) and section *Racemosae* (e.g., *C. heteroneura, C. mertensii*). Perigynia with long, scabrous beaks or teeth easily become entangled in fur or cloth (e.g., *C. comosa*), and even slight roughness can help perigynia adhere to and travel on animals. *Carex macrocephala* has long, sharp perigynium beaks that can pierce skin, and also has thick, corky bases for flotation. Many upland species have succulent perigynium bases. Ants grasp these bases and drag the perigynia off to their nests as a food source. In some ant-dispersed species, the culm continues elongating after flowering, bending over to lay the seed head on the ground where ants have easy access to the perigynia (e.g., *C. concinnoides, C. serpenticola*). In others (e.g., *C. rossii*), ants can access short pistillate spikes produced near the plant base.

In *C. pauciflora,* the perigynium base becomes compressed as the mature seeds bend downward. Slight jostling by wind or a passing animal breaks the connection to the culm and the base suddenly expands, catapulting the seed a couple of feet from the parent shoot. In addition, the long, narrow perigynia easily become entangled in fur or cloth.

Unusual timing of perigynium dispersal may support particular seed-dispersal strategies. For example, *C. obnupta* retains perigynia until they can disperse on the floods of late winter and spring. *Carex nudata* flowers and fruits early, dispersing seed as spring floods end and the disturbed riverbank soil is exposed, providing good seedling habitat.

All *Carex* perigynia stick to moist surfaces. They also mix with mud and then stick to feet, shoes, feathers, or vehicles, sometimes traveling long distances.

Figure 3. Inflorescences of *Carex spectabilis* (left) and *C. hoodii* (right) with smut fungi

Therefore, *Carex* are found as waifs along roadsides and railroads, and near the doors of botanists' homes. By this means, alkali-tolerant *C. praegracilis* and *C. duriuscula* have recently extended their ranges from the West and Midwest into northeastern North America, where suitable alkaline habitat has recently been created by salting roads in winter.

Seed production and dispersal can be reduced or eliminated by infection of the inflorescence by parasitic fungi, usually smut (Figure 3).

Seed Germination

Carex seed biology has sometimes seemed mysterious, both because the seeds can be difficult to germinate and because seedlings of most species are rarely observed in nature even where adults are common. Healthy, mature sedge meadows normally lack disturbed sites so they normally lack seedlings. *Carex* seeds can be germinated by providing environmental cues associated with good seedling habitats, which are usually moist, sunny, disturbed sites.

Except for a few species such as *C. nudata* whose seedling habitat opens up just after the seed ripens, the seeds of most species are physiologically dormant when first mature – they cannot germinate no matter what the conditions. This initial dormancy prevents germination in late summer or autumn when growing conditions may not be appropriate for seedlings. The seeds are stratified (prepared for germination) by cold and moisture during fall, winter, and early spring. If conditions are still not suitable for germination, the seeds can remain viable for years in the soil seed bank. Stratified seeds germinate when they detect the signals that indicate they are in appropriate seedling habitat. Moisture and light are required. Therefore, deeply buried seeds of most *Carex* species will not germinate. Fluctuating temperatures with warm days speed germination.

Seeds hidden in tough perigynia, such as those of *C. macrocephala, C. nebrascensis, C. obnupta,* and *C. raynoldsii*, cannot germinate until the perigyium is scarified by abrasive soil or perhaps the digestive system of a bird, allowing it to soak up moisture.

Some *Carex* have a low percentage of perigynia with viable seed, often because the plant does not have enough resources to grow seeds in all the perigynia it has produced. The fertile culms are usually initiated in late summer or fall. They elongate and flower the next spring or summer. Therefore, the number of perigynia reflects growing conditions of the previous year. If conditions are poor during flowering or seed maturation, the excess flowers abort. Unfilled seed can also result from bad weather during pollination, an isolated female clone of a dioecious plant, or other problems. Empty perigynia are fairly common in *C. hassei* and *C. utriculata.* In *Carex obnupta,* they are actually more common than perigynia with filled seed. This may be an adaptation to reduce predation in a seed dispersal strategy that leaves the hard perigynia on the culms, vulnerable to birds, all winter long.

Carex Nomenclature and Classification

Sedge Names

Carex is derived from the Latin for cutter, referring to the sharp leaf edges of many wetland sedges.

We strongly recommend learning the scientific names of sedges. Scientific names may change, but the changes are more ordered and in the long run rarer than changes in common names. Why do scientific plant names change?

Usually the issue is how broadly or narrowly we define a species, that is, how much variation we are willing to accept within one species. Sometimes, two populations are very similar but have small, more or less consistent differences. Reasonable botanists may then disagree about whether to call them members of the same species. This may be a judgement call that will never have a definitive answer, and we often alternate between two scientific names. For example, a common Cascadian sedge can be called *C. pensylvanica* var. *vespertina* if we consider it a variety of eastern *C. pensylvanica,* or *C. inops* if we consider it to be a separate species.

Sometimes names change for purely technical reasons. These changes are frustrating, but should result in future stability. A scientific name is associated with what is called the "type" specimen, which defines the name's meaning. Problems with interpreting the type can cause name changes. For example, the type specimen of *C. lanuginosa* turned out to be an immature *C. lasiocarpa,* making those two names synonyms and necessitating a new name for what had been called *C. lanuginosa*; it is now *C. pellita.* In another example, the accepted name of *C. leporina* changed to *C. ovalis* and back again because Linnaeus did not cite a type specimen when he published the name *C. leporina.*

Common names are more variable than scientific names, and in sedges most aren't really "common" anyway. Although scientific names of plants follow some rules, common names do not. Common names have a slithery quality, following an author's personal whims or morphing in ways observed in manuscripts copied by hand through the centuries: someone's one-time confusion is copied through generations of lists; unfamiliar words become familiar ones even if they don't make sense. The results can be fascinating, but confusing. For example:

The shape of the perigynia of *C. stipata* explains its common names, Sawbeak Sedge, Awl-fruited Sedge, and Awlfruit Sedge. Many people aren't familiar with awls, though, so the name has morphed to Owlfruit Sedge.

In some lists, *C. stipata* comes just before *C. tribuloides,* which is the only explanation we can see for calling *C. tribuloides* Awl-fruited Sedge, which becomes Awl-fruited Oval Sedge, an oxymoron. *Carex tribuloides* is also called Blunt Broom Sedge, Pointed Oval Sedge (a name more appropriate and more often used for *C. scoparia*), Bristlebract Sedge (hence Bristleback Sedge, easier to say but misleading), and Pennsylvania Sedge. (It does grow in Pennsylvania—and many other places—but this name is better used for *C. pensylvanica,* which is sometimes called Yellow Sedge.) Out of this confusion we have selected Tribulation Sedge as the common name for *C. tribuloides*. By the time you navigate through the key to this species, you will consider that name appropriate.

Meanwhile, *C. scirpoidea* begat Scirpoid Sedge, which begat Scirpus-like Sedge, which begat Bulrush Sedge (bulrushes are in the genus *Scirpus*), which begat Rush-like Sedge, although this *Carex* doesn't look much like a rush (*Juncus*).

Even the eternal alternation of scientific names *C. inops* ssp. *inops* and *C. pensylvanica* var. *vespertina* for our common Cascadian species is more stable than the endless instability of common names of sedges.

Most common names in this book come from the USDA Plants web site (http://plants.usda.gov/), sometimes with minor grammatical changes, but sometimes we chose a different name. There are no rules for common names!

Scientific names used in this guide follow the *Flora of North America* (Ball and Reznicek 2002) with few exceptions. We synonymize *C. heteroneura* and *C. epapillosa,* which are not distinct in the PNW or CA. We recognize *C. tiogana,* segregated from *C. capillaris*.

Carex Classification

Sedges (family Cyperaceae) are monocots. They share a common ancestor with the grasses (Poaceae) but are more closely related to rushes (Juncaceae). With about 2,000 species worldwide, *Carex* is the most diverse genus in the family.

Carex is such a large genus that we need to break it down into manageable units. Traditionally, it has been divided into two layers of subgroups: subgenera and sections. The subgenera were *Vignea, Primocarex,* and *Eucarex* (or *Carex,* to be technically correct but confusing). Recent research has divided the genus into clades, groups of related species that share a common ancestor. Some clades correspond to traditional classification, but others do not.

Our identification key divides sedges into groups using the same traits used to distinguish subgenera and sometimes sections. The relationship between the three main clades and traditional classification is discussed below, with mention of relevant identification keys in this book.

Subgenus *Vignea* Clade (Keys H, I, and J): These sedges have two stigmas per flower and lenticular achenes; short, sessile spikes; and spikes nearly always similar. You can recognize a sedge as *Vignea* even from a distance. Identifying it to species is another issue.

Core *Carex* Clade roughly corresponds to **Subgenus *Eucarex*** (Keys C, E, F, and G): These sedges usually have three stigmas and trigonous achenes, long spikes on stalks, and separate staminate and pistillate spikes, but there are exceptions to all those generalizations.

Caricoid Clade includes most of **Subgenus *Primocarex*** (Key A), the species with single, androgynous spikes. A few single-spike *Carex* do not belong in this clade but are modified *Vignea* (e.g., *C. gynocrates*) or *Eucarex* (e.g. *C. scirpoidea*). This clade also includes a few *Carex* with multiple spikes, including *Carex* section *Acrocystis* and *C. cordillerana,* as well as the genera *Kobresia, Cymophyllus, Schoenoxiphium,* and *Uncinia.* Because *Kobresia* arose within *Carex,* it is included in this book.

Each subgenus can be divided into sections. In terms of *Carex* evolution, some sections are merely assemblages of superficially similar species, but others are true clades. We find section names useful for organizing the swirling mass of sedge species names in our brains. Some are especially useful because their species are easy to recognize as members of the section, but hard to identify as species. The following three sections are the ones we find most useful.

Section *Deweyanae* (in *Vignea;* Key I2) has become more important to botanists since *C. deweyana,* previously thought to be a single common species of moist shady places, was split into *C. deweyana* itself (rare in the PNW) and the "identical triplets" *C. bolanderi, C. infirminervia,* and *C. leptopoda.* Plants that can be identified to this section but not to species can be reported as *"Carex* section *Deweyanae."*

Section *Ovales*, a.k.a. The Dreaded *Ovales,* (in *Vignea;* Key J) includes over thirty species in the PNW, most of them similar to each other. These species are all close relatives. As members of subgenus *Vignea,* they have two stigmas per flower, lenticular achenes, and short spikes. Their distinctive features are their winged perigynia arranged in gynecandrous spikes.

Section *Phacocystis* (in *Eucarex;* Key G) consists of deep-rooted wetland species with long spikes and separate staminate and pistillate spikes, but only two stigmas per flower (e.g., *C. aquatilis*). These species are important for streambank stabilization, fish habitat, forage, and basketry.

Sedge Species

The diversity of *Carex* and the great similarity among some of them raise the question, "Are these really good species?" In most cases, they are. As sedge expert Anton Reznicek has stated, in *Carex,* the gaps between species are narrow, but they're deep. The majority of "intermediates" disappear as one becomes familiar with the subtle traits that distinguish the species. *Carex* can achieve such high diversity in one area because so many of them are microhabitat specialists.

High elevation plants present a difficult classification issue. Plants isolated on different mountain ranges develop their own somewhat distinctive traits. Should the plants of each mountain range be recognized as species, subspecies, or varieties, or unnamed variants? Opinions differ. This problem is illustrated by sedges on Steens Mountain, a "sky island" in southeast Oregon that has been isolated from similar habitats for millenia. There, *C. tiogana* and *C. haydeniana* look odd, similar to but slightly different from more typical plants elsewhere. Isolation and resulting divergent evolution on Steens Mountain and other high elevation habitats may or may not warrant the naming of additional taxa.

A few sedge taxa are not fully distinct. Expect intermediates in taxon pairs we recognize as varieties or subspecies of a single species. The *C. exsiccata/C. vesicaria* pair also has many intermediates in the Cascades, although the problem seems more widespread than it is because immature individuals cannot be distinguished.

Some plants in section *Ovales* present a different problem, due in part to high rates of self-pollination. *Carex microptera* is relatively diverse and includes variation that was once recognized at the species level, as *C. festivella* and *C. limnophila.* In the related *C. pachystachya,* the alert botanist may recognize consistent variations that result from distinct lineages. The sane botanist will resist any temptation to give them names.

Hybrids

Most hard-to-identify sedges are just hard to key, not hybrids, but a diversity of real hybrids have been reported. Most hybridization occurs between species of the same section, but some intersectional hybrids occur regularly. Hybrids are not included in the identification keys. Hybrids are nearly always sterile, with deformed pollen, obviously deformed anthers that do not open, and perigynia that do not contain firm, well-filled achenes. A rhizomatous hybrid may form an extensive, long-lived stand, making hybridization seem more common than it is. Despite the low fertility of most hybrids, some species apparently have arisen from hybridization events, particularly in the section *Phacocystis*. Also, gene flow among fairly distinct species occurs with some regularity in certain species complexes, e.g., the one including *C. gynodynama* and *C. mendocinensis*.

Carex Morphology

Rhizomes. The most consistently perennial part of a *Carex* is the *rhizome*, an underground stem that produces roots and shoots. The rhizome's nodes and scales (very reduced leaves) distinguish it from the roots. Each species' rhizomes run at a characteristic depth below the soil surface. Some *Carex* have vertical as well as horizontal rhizomes, especially species that live in moving sand (e.g., *C. macrocephala*) or areas of sedimentation in floodplains (e.g., *C. aquatilis*). *Carex* with vertical rhizomes may form large raised clumps called *tussocks* (e.g., *C. cusickii*). In some species, the rhizome is so short that it is hardly detectable and all the shoots arise in a dense cluster. We call sedges with this "bunchgrass" growth form *cespitose*, while plants with the rhizome apparent between the shoots we call *rhizomatous*. The synonymous terms *loosely cespitose* and *short-rhizomatous* are used for plants that form clumps from short rhizomes. Rhizomatous sedges may produce shoots singly, at a distance that is characteristic of the species (see *C. pansa*). They may have long rhizomes that produce clusters of shoots, especially where resources are concentrated (e.g., *C. obnupta*). The difference between rhizomatous and cespitose growth form is basic to *Carex* identification.

Roots. The reader will be pleased to hear that few root features are used in identification. *Carex limosa* and allies have a dense, yellow, felty covering on the roots. *Carex spectabilis* and relatives may have a similar white covering.

Shoots. Sedges have vegetative shoots, which produce only leaves, and fertile shoots, which produce inflorescences.

Vegetative shoots. In most *Carex*, the vegetative shoot is a *pseudoculm*, consisting of overlapping leaf sheaths and their respective blades, with the true stem reduced to a tiny nubbin. Peel off the leaves with their sheaths; when you remove the last one, nothing is left but a tiny, basal stub of stem. A few sedges have *true vegetative culms*, with elongated triangular stems and separated nodes (see *C. chordorrhiza*). All members of section *Ovales* have true vegetative culms, though they may be short, and they are found in a few other species such as *C. atherodes* and *C. sheldonii*. True vegetative culms may form *stolons*, horizontal, above-ground stems that give rise to more shoots and roots (e.g., *C. leporina, C. limosa, C. chordorrhiza*).

Fertile shoots. When a shoot begins growing on a rhizome, it is hidden by a few bladeless leaf sheaths that function like bud scales. A fertile shoot that starts growth and produces an inflorescence in the same year has these bladeless sheaths at the base and is called *aphyllopodic* (*a* = lacking; *phyl* = leaf blade; *pod* = foot or base: *aphyllopodic* = lacking leaf blades at base). Fertile shoots produced from last year's vegetative shoots have remnants of last year's decayed leaf blades at the base and are called *phyllopodic.*

The buds that will develop into flowers are usually initiated late in the year, then become dormant. The next spring, these fertile shoots elongate and flower. Therefore the number of flowers and seeds produced by a *Carex* plant reflects environmental conditions in the previous year. If the shoot initiated more flowers than can be supported by conditions during the flowering season, the excess flowers may abort.

Leaves. The *Carex* leaf consists of a sheath, which surrounds the culm (if there is one), and a linear blade, with a *ligule* at the junction of sheath and blade. The ligule is a flap of tissue that may reduce flow of water down the blade into the sheath, preventing fungus growth. In sedges, the ligule is mostly fused to the blade, leaving only a narrow free portion.

The *leaf sheath front* has useful traits for identification, but where is the front? Consider the leaf blade to be Superman's cape. The sheath is his shirt, and the front is the part on his chest. The leaf sheath front is usually white *hyaline* (pale, smooth, translucent, and veinless, becoming white when dry), but it may be red-dotted (speckled), coppery, green and veined, or have other odd traits. Veined leaf sheath fronts may be *fibrillose,* disintegrating to leave only the fibers, which may form a pattern called *ladder-fibrillose,* with a vertical central fiber from which lateral fibers arise (see *C. obnupta, C. pellita*). The leaf sheath front may be *cross-corrugated* (= cross-rugulose) with a regular pattern of horizontal wrinkles, waves, or puckers (see *C. densa*). The top of the leaf sheath front is the *mouth,* which may have a characteristic color or shape. The mouth may be elongated and fused with the ligule to form a sleeve-like *contraligule.* In leaves that arise under water, relatively large air pockets may form in the backs and sides of the sheaths, and the sheaths are strengthened by numerous cross-walls between the longitudinal veins, forming a pattern like brick-work (see *C. utriculata*); such sheaths are called *septose-nodulose.* In a few sedges of dry uplands, the sheaths are tough, brown, and relatively resistant to decay (*C. filifolia, C. nardina*).

Flowers. *Carex* have separate *staminate* (male) and *pistillate* (female) flowers. Each staminate flower consists of one to three stamens. Immediately below the staminate flower is a bract called the *staminate scale.* In other words, the staminate scale *subtends* the staminate flower. The anthers usually fall off early in development, but the white, thread-like filaments persist.

Sedge Parts

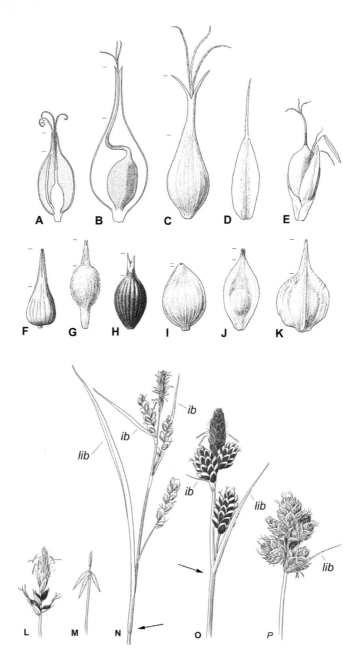

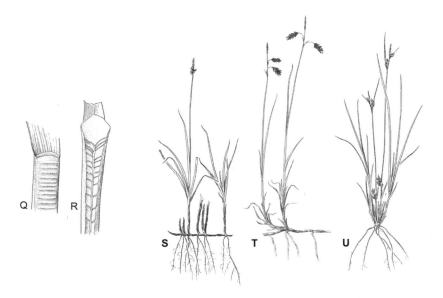

A – B: dissected perigynia. A – showing rachilla. B –showing persistent style, inflated perigynium.

C – Inflated perigynium with 2 teeth, 3 stigmas. D – pistillate scale with awn. E – Open perigynium (*Kobresia*) showing achene and (behind bract) male flower.

F – K: perigynia. Arrows show tip and base of beak. F – lance-triangular. G – pubescent, with long stipe. H – with bidentate beak (2 teeth). I – beakless. J – winged. K – winged.

L – P: inflorescences; ib = inflorescence bracts. lib = lowest inflorescence bract.

    L – M: single androgynous spikes. L – dark perigynia and pale pistillate scales. M –with strongly reflexed perigynia and deciduous scales. N – O: terminal spike staminate, lateral spikes pistillate; arrow indicates node at base of sheath of lowest inflorescence bract. N – lowest inflorescence bract with well-developed (inflated) sheath O – lowest inflorescence bract sheathless. P – dense head with crowded gynecandrous spikes, bristle-like inflorescence bract.

Q – cross-corrugated leaf sheath front. R – ladder-fibrillose leaf sheath front.

S – rhizomatous plant with aphyllopodic shoots (lowest leaves reduced to bladeless sheaths); shoots arising singly. T – rhizomatous plant with phyllopodic shoots (with remnants of last year's blades near the base). U – cespitose plant with basal pistillate spikes.

Each pistillate flower consists of an ovary, a style, and two or three stigmas (four in *C. concinnoides*). The ovary develops into an *achene,* a dry, one-seeded fruit. If the flower has two stigmas, the achene is *lenticular* (two-sided). If there are three stigmas, the achene is *trigonous* (three-sided). (In *C. concinnoides,* our only four-stigma sedge, the achene is ± globose with a ± four-sided base.)

In most species, the style is *deciduous.* It breaks off of the achene apex after pollination, leaving a little stub. In an immature deciduous style, a color change indicates where the style will beak. In a few species, the style is *persistent,* remaining on the achene and often becoming twisted as the achene grows (Figure 4). A young persistent style is usually uniform in color.

Figure 4. Persistent styles on achenes of *C. utriculata*

Perigynia. The *perigynium* is a modified bract that encloses the pistillate flower. The perigynium is subtended by a small bract called the *pistillate scale.* The midrib of the scale may extend beyond the flat part to form an *awn*.

The perigynium consists of the *body*, the part around the achene, and the *beak*, the narrowed portion that surrounds the style. The beak tip often has two small projections called *teeth* (see long teeth in *C. comosa*). In some species, the base of the body is developed into a short, stalk-like *stipe*. The perigynium body may have flat edges called *wings* (e.g., *Carex* section *Ovales*). Nearly all perigynia have two *ribs,* fibrous vascular bundles usually located at the edge of the perigynia. Most also have obvious *veins*, which are smaller vascular bundles with their associated fibers, on the dorsal and ventral surfaces.

Which side of a perigynium is which? The *ventral* side is the side toward the axis of its spike; it is usually somewhat concave. The *dorsal* side is away from the axis of its spike, toward the pistillate scale. Think of the perigynium as a squirrel running up a tree, its belly (ventral side) toward the tree trunk.

In some species, the perigynium encloses a *rachilla*, a vestigial flower stalk. Most *Kobresia* perigynia, which are open on one side, enclose both a pistillate flower and a stalked staminate flower.

Spikes. The basic unit of the sedge inflorescence is the *spike,* which consists of a central axis and, connected directly to it, staminate flowers and/or perigynia and their subtending scales. An inflorescence may consist of one or several spikes. A spike may be *sessile* (connected directly to the culm or side branch) or stalked (on a *peduncle*). It may be *staminate* (with only male flowers), *pistillate* (with only female flowers), *androgynous* (with male flowers at the tip and female at the base), *gynecandrous* (with female flowers at the tip and male flowers at the base), or mixed (with an irregular arrangement of male and female flowers). In a few of our species (*C. cusickii, C. densa, C. divulsa,* and *C. vulpinoidea*) the inflorescence is branched, with some spikes attached to side branches. If there is just one very short side branch, the only evidence of it may be that the lowest infloresence node looks like it produces two sessile spikes.

Inflorescences. *Carex* plants are *monoecious* (with the male and female flowers on the same plant) or, less often, *dioecious* (with male and female flowers on different plants). A few *Carex* have separate male and female inflorescences on different shoots of the same plant (*C. kobomugi, C. macrocephala,* some *C. serpenticola*). For these plants, botanists have created the nifty term *paradioecious,* which you will probably never see again.

Inflorescence bracts. In general, each spike grows from the axil of an *inflorescence bract.* Inflorescence bracts may be leaf-like (e.g., *Carex aquatilis*) or bristle-like (e.g., *C. pachystachya*). In sedges, "bristle-like" does not suggest stiffness, merely that the bract consists of little more than the midrib and therefore is not leaf-like. The lowest inflorescence bract may be virtually sheathless (e.g., *C. obnupta*) or may have a well-developed sheath. If the well-developed sheath fits loosely around the culm, it is called *inflated* (e.g., *C. aurea, C. hendersonii, C. luzulina*).

Collecting Sedges

Field identification of sedges can be difficult so it is sometimes useful to collect the plant for later study with a microscope and good light. Plant collections are also essential for documenting the distribution of species, especially for newly introduced species or for range extensions of native species.

"Roots, Fruits, and Shoots!" – This is the sedge collector's commandment. Collect them all. Take notes on habitat and growth form (rhizomatous, cespitose, etc.). Underground parts may be essential for identification. In many groups, identification requires mature perigynia, which means they should fall off relatively easily when you rub the inflorescence. Collect several mature shoots, being careful to collect only one species at a time and place each species in its own bag, especially if it is mature enough that the perigynia are falling.

For native plants, collections should only be made from populations that are large enough to sustain the loss of individuals. A rough rule of thumb is the "one in twenty rule" which allows no more than 5% of the culms to be collected. Collections should not be made from very small populations or from populations of very rare native species. Some plant species can be adequately documented with good quality photographs, but a specimen is often necessary to confirm or refute a *Carex* identification. If collection seems necessary but the popuation is small, minimize harm to a population by collecting one or two fertile shoots and leaving the perennial rhizome undisturbed.

If the collection is donated to an herbarium, it should be pressed flat in newspaper and dried. Collection information must include the date and place of collection (preferably with latitude/longitude or UTM determined in the field with a GPS unit), and collector's name. Information on habitat and population size can be useful.

Significant collections should be donated, with collection data, to the major regional herbaria at Oregon State University, University of Washington, or Washington State University. It is a good idea to donate duplicates to herbaria of local land management agencies and smaller colleges and universities.

Tools for Collecting and Identifying Sedges

Sedge knives. Serrated kitchen knives (plural) with stiff blades, available in thrift shops, to cut toughs roots and rhizomes. Tie colorful flagging around the handles so it will take longer to lose them.

Plant press and newspapers. Put the sedge inside a folded newspaper. Number the sheet or otherwise identify where it came from. Press it flat and dry it.

Plastic bags. Many of them, so that each sedge can be bagged and labeled separately in the field. Otherwise perigynia mingle as they fall off.

Hand lens. 10X – 16X.

Rulers. Six inch, with metric scale. Buy many—you'll lose these, too.

Dissecting microscope (optional but very helpful). Perhaps at your workplace, a local school, or an herbarium.

Reticle, a small metric scale for use with a microscope. Some scopes have a reticle in the eyepiece already. Alternatively, you can lay your reticle right on the plant. Get one that measures 5 or 10 mm in tenths. For example, Bausch & Lomb metric scale – part # 813438.

Probes and fine forceps.

Pads of little sticky notes. Put perigynia on the adhesive strip so they won't fly away.

Fragment packets, to package loose perigynia. Make your own out of paper scraps, or buy ones made from acid-free paper from companies that provide supplies for herbaria. We like the 1-11/16 by 2-1/4 inch size.

How to Use the Identification Key

Read the first pair of matched descriptive statements (leads); these have the same number. Decide which lead fits your plant best. Go to the next pair of leads below the one that fits your plant best, or follow instructions to a subkey. When you get to a sedge name, stop and check the description and pictures in the species account to determine if you have correctly identified your plant.

If you can't decide which of the two paired leads fits your plant best, go both ways, trying first one and then the other. You may find possible identifications, which you can check among the species accounts, or you may find some pairs of leads that are so inappropriate to your specimen that you can eliminate that route through the key. Sometimes a single species may key out on both sides of a key lead pair. A majority of the species key out more than once, some species several times.

For unfamiliar vocabulary, see the Glossary, p. 415. Also, the on-line glossary created by Tim Jones (http://www.herbarium.lsu.edu/keys/gloss/glossary.htm) uses many *Carex* examples.

Before You Key

Examine your plant and figure out its basic traits, things you're going to need to know to key it. Is the plant rhizomatous or cespitose? What habitat does it occupy? What is the arrangement of staminate and pistillate flowers? Are the stigmas 2 (and achenes lenticular) or 3 (and achenes trigonous)? Is the inflorescence a single spike (Key A)? Does the inflorescence consist of short spikes that are all similar? If so, the plant is probably in subgenus Vignea, Keys H, I, and J. Does the plant have unusual traits such as hairy perigynia or leaves, cross-corrugated leaf sheath fronts, or unusually large perigynia?

Common Questions

Two stigmas or three? By the time a plant has perigynia mature enough to identify, the stigmas may have fallen. Two-stigma flowers produce two-sided (lenticular) achenes. Three-stigma flowers produce three-sided (trigonous) achenes. Usually, but not always, the perigynium shape follows the achene shape; when in doubt, dissect out the achene. In the field, you may be able to judge achene shape by squeezing and rolling the perigynium between your fingers.

Where are the staminate (male) flowers? Sedges are classified by the relative position of male and female flowers on the spikes, but the anthers fall long before the perigynia mature. Fortunately, the filaments remain. Bend back a scale and look for the white, thread-like filaments. Pistillate (female) flowers produce perigynia. To find out if a spike is gynecandrous (girls on top) or androgynous (boys on top), peel back the lowest two or three scales (bracts) to see what's underneath. Examining spike tips may help, but be careful because empty-looking bracts might hold female flowers that have not matured.

Which perigynia to measure? Perigynia at the spike base are often warped and too wide. Those at the tip are too narrow and sometimes too long. The ones in the middle third are just right. This is especially important on short spikes.

How to keep those pesky perigynia from flying away? Put them on the adhesive portion of a sticky note.

How to Measure Sedge Parts

Achene length includes the stipe (if any) at the base but excludes the style or the stub where the style broke off.

Beak length is the distance from the tip of the perigynium to the inflection point of the margin, the point where the margin turns toward the beak (Figure 5).

Inflorescence length is measured from the bottom of the sheath of the lowest inflorescence bract to the tip of the uppermost spike.

Ligules: The ligule is largely fused to the blade. Measure the distance from the "corner" where the edge of the blade joins the sheath, to the place where the ligule crosses the midrib of the blade.

Perigynium thickness: Moisten your finger tip or a dissecting tool and stick the perigynium to it; that way you can hold it on edge under the microscope.

Figure 5. Perigynium of *Carex pachycarpa* showing beak

How to Interpret Photographs

Most photographs in the species accounts show live plants in the field. Herbarium specimens may appear faded, deflated, or wrinkled. Perigynia change color with age, starting green and becoming straw-colored, or starting green or tan and becoming dark green or brown. However, dark reddish or blackish pigmentation may remain stable.

How to Use the *Ovales* Key

Carex section *Ovales* is the most difficult group, with over thirty species in the PNW. Many of them have evolved relatively recently, perhaps since the last glacial retreat 10,000 years ago. Distinguishing them is challenging but the following guidelines will help keep you on track.

Perigynium measurements: Take measurements from mature perigynia in the middle third of the spike. Examine several perigynia and take measurements from the larger ones that are normal in shape.

Beaks: A large group of western *Ovales* is characterized by having perigynium beak tips that are slender, dark brown, unwinged, parallel-sided, and entire, like small, slightly flattened tubes; see pictures of *C. microptera* and *C. pachystachya* perigynia. Usually the unwinged beak tip is at least 0.4 mm long, but in species that have small perigynia, like *C. subfusca,* it may be shorter. The trait is easily overlooked but is distinctive once seen. The contrasting trait is to have the beak winged (flat margined, not parallel-sided) and cilliate-serrulate to the tip (or very nearly to the tip).

Inflorescence shape: Plants that key as having a dense head have really, really, dense, almost round heads. The spikes are somewhat difficult to distinguish. Plants with more oval inflorescences, which in any other group might be considered dense, key as having somewhat elongated inflorescences. When in doubt, key the plant as having first dense, then elongate inflorescences.

Late season shoots: Shoots that grow up late in the season, after most of the shoots flower, may be unusually elongated or dense, with abnormally long bracts and odd-shaped perigynia. They may be tricky to identify.

Leaf sheaths: In a dried specimen, a white-hyaline leaf sheath front is whitish. In a fresh plant, it is usually transparent and hard to see. You may look right through it and see the green, veined back of the next leaf sheath inside it. Tear the fresh leaf sheath front lengthwise; a white or silvery margin of the tear indicates that the sheath is white-hyaline.

Wing: As used here, the wing is the entire flat margin of the perigynium body, including but not limited to the area beyond the outermost obvious vein.

Abbreviations and Symbols

mt = mountain
N, S, E, W = north, south, east, west
PNW = Pacific Northwest
♂ = male or staminate. ♀ = female or pistillate.
± = more or less; approximately
< = smaller than, less than, fewer than
 > = larger than, greater than, more than

States and provinces are abbreviated using standard postal codes:
Canadian Provinces: AB Alberta, BC British Columbia, MB Manitoba, NB New Brunswick, LB Labrador, NF Newfoundland, NT Northwest Territories, NS Nova Scotia, NU Nunavut, ON Ontario, PE Prince Edward Island, QC Quebec, SK Saskatchewan, YT Yukon

U.S. states: AK Alaska, AL Alabama, AR Arkansas, AZ Arizona, CA California, CO Colorado, CT Connecticut, DE Delaware, FL Florida, GA Georgia, HI Hawaii, IA Iowa, ID Idaho, IL Illinois, IN Indiana, KS Kansas, KY Kentucky, LA Louisiana, MA Massachusetts, MD Maryland, ME Maine, MI Michigan, MN Minnesota, MO Missouri, MS Mississippi, MT Montana, NC North Carolina, ND North Dakota, NE Nebraska, NH New Hampshire, NJ New Jersey, NM New Mexico, NV Nevada, NY New York, OH Ohio, OK Oklahoma, OR Oregon, PA Pennsylvania, RI Rhode Island, SC South Carolina, SD South Dakota, TN Tennessee, TX Texas, UT Utah, VA Virginia, VT Vermont, WA Washington, WI Wisconsin, WV West Virginia, WY Wyoming

Updates to the key will be posted at http://www.carexworkinggroup.com

Identification Key to Sedges of the Pacific Northwest

1a. Spike one per culm .... Key A, p. 38

1b. Spikes two or more per culm

  2a. All spikes entirely ♂; no perigynia present .... Key B, p. 41

  2b. Some or all spikes with ♀ flowers; perigynia present

    3a. Perigynia hairy .... Key C, p. 42

    3b. Perigynia glabrous (may have cilia on beak margins)

      4a. Stigmas 3; achenes 3-sided in cross section

        5a. Perigynia open on one side, exposing the achene; plant short, densely cespitose; habitat high elevation bogs, meadows, and rock fields .... *Kobresia,* Key D, p. 44

        5b. Perigynia closed, hiding the achene; habit and habitat various

          6a. Styles persistent; perigynia usually inflated .... Key E, p. 44

          6b. Styles deciduous; perigynia usually tight against the achenes. ....Key F, p. 45

      4b. Stigmas 2; achenes 2-sided in cross section

        7a. Some spikes entirely ♂; spikes often > 1.5 cm long

          8a. Foliage normally orange-brown; ♀ scales white hyaline throughout except midrib; plants of disturbed places, rarely escaping cultivation W of Cascades ..... *C. buchananii,* p. 410

          8b. Foliage normally green or glaucous; ♀ scales not hyaline throughout; habitat various

            9a. Perigynia winged; habitat disturbed sandy sites in Portland area ....*C. arenaria*

            9b. Perigynia not winged; habitat various..........Table G, p. 51

        7b. Every spike with at least some ♀ flowers; spikes usually < 1.5 cm long

          10a. Perigynia unwinged (may be flat-edged on beak)

            11a. Spikes androgynous, ♀, or with ♂ and ♀ flowers mixed . ... Key H, p. 55

            11b. Spikes gynecandrous, or occasionally with terminal spike gynecandrous and lateral spikes ♀ .... Key I, p. 60

          10b. Perigynia winged (flat-edged) at least on distal part of body

            12a. Plants cespitose; spikes gynecandrous or pistillate ....Key J, p. 62

            12b. Plants rhizomatous; spikes staminate, androgynous, or pistillate

              13a. Perigynia with broad wings especially at base of beak; ♀ scales usually longer than the perigynia; introduced in sandy soils in Portland area .... *C. arenaria*

              13b. Perigynia with a narrow wing about 0.1 mm wide; ♀ scales shorter than or about as long as the perigynia; native to dry pine savanna and grassland in E WA .... *C. siccata*

Key A: Inflorescence consisting of a single spike

1a. Spike entirely ♂; no perigynia present .... Key A1, p. 38
1b. Spike entirely or partially ♀; perigynia present
  2a. Perigynia pubescent or puberulent, at least at base of beak .. Key A2, p. 39
  2b. Perigynia glabrous (sometimes minutely papillose; sometimes serrulate
    on the edges of the beak)
    3a. Perigynia linear to narrowly elliptic, length (including stipe) > 3 times
      the width, AND beakless or tapering gradually to a poorly defined beak
      .... Key A3, p. 39
    3b. Perigynia elliptic to ovate, length < 3 times the width, and/or tapering
      to a distinct beak
      4a. Stigmas 2; achenes lenticular .... Key A4, p. 40
      4b. Stigmas 3; achenes trigonous
        5a. Perigynia open on one side, exposing the mature achene;
          remnants of previous year's leaf sheaths dark brown, shiny, and
          abundant at base of plant; leaves to 0.5 mm wide ....
          ....*Kobresia myosuroides*
        5b. Perigynia closed; remnants of previous year's leaf sheaths absent
          or various including dark brown, shiny, and abundant at base of
          plant; leaves 0.2-6.0 mm wide
          6a. Perigynia beakless, blunt, green to tan .... Key A5, p. 40
          6b. Perigynia beaked, sometimes tapered gradually to the beak,
            usually tan to brown
            7a. Lowest inflorescence bract leaf-like, much longer than
              inflorescence, much > 2 cm long............*Carex cordillerana*
            7b. Lowest inflorescence bract not leaf-like, usually < 2 cm
              long
              8a. Spike ♀ or gynecandrous .... *C. idahoa*
              8b. Spike androgynous .... Key A6, p. 41

Key A1: Spike 1 (-4), entirely ♂

1a. Leaves 0.4-0.9 mm wide .... *C. gynocrates*
1b. Leaves 1-5 mm wide
  2a. Some ♂ scales minutely ciliate on edges, at least at tip
    3a. Plants in SW OR
      4a. Flowering April-early May; ligules wider than long . *C. serpenticola*
      4b. Flowering June-August; ligules longer than wide
        5a. Leaves 2.8-4 mm wide; plants +/- cespitose; substrate serpentine
          .... *C. scabriuscula*
        5b. Leaves <3 mm wide; plants often rhizomatous; substrate
          serpentine or not.................. *C. scirpoidea* ssp. *pseudoscirpoidea*
    3b. Plants elsewhere in the PNW .... *C. scirpoidea* (with 3 subspecies)
      6a. Fertile culms phyllopodic; plants ± rhizomatous.............................
      ....*C. s.* ssp. *pseudoscirpoidea*
      6b. Fertile culms aphyllopodic; plants ± cespitose

> 7a. Habitat wet .... *C. s.* ssp. *stenochlaena*
> 7b. Habitat mesic to dry .... *C. s.* ssp. *scirpoidea*

2b. ♂ scales entire to erose, not ciliate
> 8a. ♂ scales entirely brown; range Olympic Peninsula......*C. anthoxanthea*
> 8b. ♂ scales with distinct hyaline margin; range SW OR or E of Cascades
> > 9a. Mouth of leaf sheath minutely ciliate; culm 10-20 cm long; range SW OR ....*C. serpenticola*
> > 9b. Mouth of leaf sheath entire; culm 15-45 cm long; range E of Cascades ....*C. idahoa*

Key A2: Spike 1 (-3); perigynia pubescent at least at base of beak

1a. Leaves involute, 0.3-0.7 mm wide; spike androgynous...................................
....*C. filifolia* var. *filifolia*
1b. Leaves flat or V-shaped in cross section, 2.8-4 mm wide; spike ♀ or with some ♂ flowers mixed among the ♀ ones
> 2a. Plants occurring in SW OR
> > 3a. Perigynia green to tan at maturity, round in cross section, thick enough to push the ♀ scales away from the axis of the spike, the spike therefore with a jagged outline; perigynium bases tapered; flowering April to May; range Josephine and Curry cos., OR, and Del Norte Co., CA; substrate serpentine ....*C. serpenticola*
> > 3b. Perigynia black, brown, or tan at maturity, thinner and not pushing the ♀ scales away from the axis of the spike, the spike therefore with a smooth, cylindrical shape; perigynium bases tapered or truncate; flowering June to August; range and substrate various, including serpentine
> > > 4a. ♀ scales narrower than the mature perigynia; perigynia ± flat ....
> > > .... *C. scabriuscula*
> > > 4b. ♀ scales wider than or equal to the perigynia; perigynia trigonous .
> > > .... *C. scirpoidea* ssp. *pseudoscirpoidea*
> 2b. Plants occurring elsewhere in the PNW..
> ....*C. scirpoidea* (with 3 subspecies)
> > 5a. Perigynia lanceolate to elliptic, (2.8-)3-4+ mm long, mostly > 2.5 times as long as wide; ♀ scales lanceolate, 3.5 mm long........................
> > ....*C. s.* ssp. *stenochlaena*
> > 5b. Perigynia ovate, (1.3-)2-2.8(-3) mm long, < 2.5 times as long as wide; ♀ scales ovate, 2.5-3 mm long
> > > 6a. Fertile culms phyllopodic; plant ± rhizomatous ....
> > > ....*C. s.* ssp. *pseudoscirpoidea*
> > > 6b. Fertile culms aphyllopodic; plant ± cespitose . *C. s.* ssp. *scirpoidea*

Key A3: Spike 1; perigynia linear to narrowly elliptic, 3–9 times as long as wide

1a. Perigynia virtually beakless, ± rounded at the tip..........................*C. leptalea*
1b. Perigynia beaked

2a. Plants densely cespitose
    3a. Perigynia 5-9 times as long as wide, with faint fine veins that are longer than the achenes .... *C. circinata*
    3b. Perigynia 3-4.5 times as long as wide, with two marginal ribs but otherwise veinless or with veins only near the base
        4a. Leaf bases from previous year not persistent; perigynia spreading to reflexed at maturity, with margins and beaks smooth .. *C. micropoda*
        4b. Leaf bases from previous year persistent; perigynia ascending at maturity, with margins and beaks ± serrulate .... *C. nardina*
2b. Plants rhizomatous to loosely cespitose
    5a. Perigynia veinless but with two lateral ribs; leaves (1.5-)2-4 mm wide ....*C. nigricans*
    5b. Perigynia with veins on the faces as well as two lateral ribs; leaves 0.5-2.5 mm wide
        6a. Perigynia 3-4.3 mm long .... *C. anthoxanthea*
        6b. Perigynia 4.5-8 mm long
            7a. Perigynia 5.9-8 mm long, reflexed at maturity; ♀ scales falling before the perigynia .... *C. pauciflora*
            7b. Perigynia 4.5-6 mm long, ascending; ♀ scales persistent .... *C. circinata*

Key A4: Spike 1; perigynia elliptic to ovate; stigmas 2; achenes lenticular
1a. Plants densely cespitose
    2a. Perigynia ovate to orbicular, sessile .... *C. capitata*
    2b. Perigynia elliptical, stipitate
        3a. Leaf bases from previous year not persistent; perigynia spreading to reflexed at maturity, with margins and beaks smooth ..... *C. micropoda*
        3b. Leaf bases from previous year persistent; perigynia ascending at maturity, with margins and beaks ± serrulate .... *C. nardina*
1b. Plants rhizomatous or stoloniferous
    4a. Perigynia reflexed when mature; plant rhizomatous .... *C. gynocrates*
    4b.Perigynia ascending when mature; plant stoloniferous.......*C. chordorrhiza*

Key A5. Spike 1; perigynia elliptic to ovate, nearly beakless, stigmas 3
1a. Lower ♀ scales leaflike, 10+ mm, much longer than the perigynia; leaves 1.5-6 mm wide .... *C. cordillerana*
1b. All ♀ scales scale-like, not or little longer than the perigynia; leaves 0.4-3.5 mm wide
    2a. Leaves and culms delicate, leaves 0.4-1.3 mm wide; perigynia 2.4-4.9 (-5.4) mm long; habitat wetlands .... *C. leptalea*
    2b. Leaves and culms tough; perigynia (3.6-)4-8.5 mm long; habitat uplands
        3a. Culms sharply triangular; leaves >/= culms; plants loosely cespitose to short-rhizomatous; widespread .... *C. geyeri*
        3b. Culms +/- cylindric; leaves shorter than culms; plants densely cespitose; SW OR .... *C. multicaulis*

Key A6: Spike 1, androgynous; perigynia elliptic to ovate; stigmas 3

1a. Perigynia 1-3, 5-7+ mm long; habitat montane .... *C. geyeri*
1b. Perigynia 5-20, 2-7 mm long; habitat subalpine to alpine
  2a. Plants densely cespitose
    3a. Leaf bases from previous year not persistent; perigynia spreading to reflexed at maturity, with margins and beaks smooth ..... *C. micropoda*
    3b. Leaf bases from previous year persistent; perigynia ascending at maturity, with margins and beaks ± serrulate .... *C. nardina*
  2b. Plants loosely cespitose to long rhizomatous, sometimes with shoots arising in clusters from long rhizomes
    4a. Perigynia finely veined on the faces, with veins extending to above the achene .... *C. obtusata*
    4b. Perigynia lacking veins on the faces or with veins present at the base but not extending to above the achene, often with two marginal ribs
      5a. Perigynia 2.5-3.9 mm long, only somewhat larger than the achene; spikes narrower, often 4-5 mm wide, 2.5-3 times as long as wide .... *C. subnigricans*
      5b. Perigynia (3.5-)4-7 mm long, much larger than the achene; spikes oval, often 6-12 mm wide, 1.2-2.5 times as long as wide
        6a. ♀ scales with 3-5 veins and with wide, whitish central portion; perigynia broadly elliptic to orbicular with short veins .... *C. breweri*
        6b. ♀ scales with 1 vein and with narrow, yellow-brown to light brown central portion; perigynia elliptic, veinless......................... .... *C. engelmannii*

Key B: Inflorescences normally entirely ♂

1a. Terminal spike much longer than the lateral spikes; lateral spikes usually 1-3 and very easy to distinguish .... Key A1, lead 3, p. 38
1b. Terminal spike similar in length to lateral spikes; lateral spikes many, all or most of them crowded and difficult to distinguish
  2a. Widest leaves 4-8 mm wide; habitat sandy soils, near the coast or along the lower Willamette and Columbia Rivers .... Key F2, p. 46
  2b. Widest leaves 0.3-4 mm wide; habitats various, including sandy soils
    3a. Widest leaves 0.3-0.7(-1) mm wide; rhizomes not more than 1 mm wide, mostly without persistent vestigial leaf sheaths; leaf sheaths at base of culms and on rhizomes not or scarcely fibrous; plants delicate, growing in moss mats in bogs and fens .... *C. gynocrates*
    3b. Widest leaves 1+ mm wide; rhizomes mostly > (0.8-)1 mm wide, covered with persistent vestigial leaf sheaths; leaf sheaths at base of culms and on rhizomes disintegrating into coarse persistent fibers; plants not delicate; habitats various but usually alkaline or sandy, not growing in moss mats
      4a. Rhizomes < 2.1 mm thick, brown; leaf blades flat with involute tips or involute throughout

      5a. Inflorescence 1.5-2.6 cm long; anther awns 0.2-0.4 mm long; common E of Cascades .... *C. douglasii*

      5b. Inflorescence 0.9-1.3 cm long; anther awns < 0.1 mm long; rare and local E of Cascades, last collected there in 1919 ....*C. duriuscula*

   4b. Rhizomes > 2 mm thick, brown to blackish; leaf blades flat or V-shaped in cross section

      6a. Habitat coastal sands; longest anther awns 0.2-0.4 mm long; filaments generally hidden behind the ♂ scales .... *C. pansa*

      6b. Habitat various but not coastal sands; longest anther awns 0.1-0.2 mm long; filaments generally exserted from the ♂ scales

         7a. Plant bases dark brown to black; anther awns usually hairy (at 20X), the tip tapered or obtuse; habitat meadows, prairies, roadsides, seasonally wet but dry at rhizome depth in summer .... *C. praegracilis*

         7b. Plant bases pale to brown; anther awns glabrous (at 20X), the tip obtuse; plant bases gray brown to medium or dark brown; habitat wet meadows, marshes, wet stream banks, where the soil is wet at rhizome depth all year .... *C. simulata*

Key C: Perigynia hairy

1a. Inflorescence entirely ♀ or with ♂ flowers scattered in mainly ♀ spikes........ .... Key A2, p. 39

1b. Inflorescence with terminal spike ♂ (or rarely gynecandrous)

   2a. Lateral ♀ spikes 1-3, < 1.5 cm long; inflorescence (excluding any basal spike) usually < 4(-6) cm long .... Key C1, 42

   2b. Lateral ♀ spikes 1-many, usually > 2 cm long; inflorescence > 4 cm long

      3a. Leaf sheath fronts pubescent at least toward top; leaf blades usually hairy .... Key C2, p. 43

      3b. Leaf sheath fronts glabrous; leaf blades glabrous .... Key C3, p. 43

Key C1: Perigynia hairy; small upland plants with short inflorescences, and short spikes; section *Acrocystis* except as noted

1a. Plants with basal spikes nestled among the plant bases, as well as spikes at the top of the culms; lowest inflorescence bract longer than the inflorescence (except in strictly coastal *C. brevicaulis*)

   2a. Perigynia several-veined to at least mid-body; leaves glaucous (occasionally green); lower surface of leaves densely papillose (at 20X); range SW OR .... *C. brainerdii*

   2b. Perigynia veinless except for 2 ribs (or somewhat veined only at the base); leaves green (occasionally glaucous); lower surface of leaves smooth or sparsely papillose; range various, including SW OR

      3a. Habitat sandy soils at the immediate coast; lowest inflorescence bract often shorter than the inflorescence; leaves strongly arching ....*C. brevicaulis*

3b. Habitat various, including non-coastal sites and coastal headlands but
not coastal sands; lowest inflorescence bract equaling or longer than the
inflorescence; leaves not strongly arching
    4a. Perigynia from non-basal spikes with beaks 0.4-0.8 mm; culms
relatively smooth except near inflorescence; habitat montane ....
....*C. deflexa* var. *boottii*
    4b. Perigynia from non-basal spikes with beaks 0.9-1.6 mm; culms
definitely scabrous; habitat various, including montane........*C. rossii*
1b. Plants lacking basal spikes; lowest inflorescence bract shorter than the
inflorescence
    5a. Stigmas 4, warty (at 15X); achenes quadrangular at base; section
Clandestinae ....*C. concinnoides*
    5b. Stigmas 3, plumose (at 15X); achenes trigonous at base
        6a. ♀ scales dark purple to black; substrate serpentine; range SW OR ......
....*C. serpenticola*
        6b. ♀ scales brown or reddish brown; substrate non-serpentine; wide-
spread
            7a. ♂ spikes 1-2.5 cm long; lowest inflorescence bract > 1 cm long;
perigynium beak 0.5-1.5 mm long .... *C. inops* ssp. *inops*
            7b. ♂ spikes 0.3-0.7 cm long; lowest inflorescence bract < 1 cm long;
perigynium beak to 0.5 mm long; section *Clandestinae* . *C. concinna*

Key C2: Perigynia hairy; leaf sheath fronts pubescent at least toward top; leaf blades usually hairy

1a. Perigynium beak teeth < 0.8 mm; perigynia 2.5-5.3 mm long; style
deciduous; range widespread
    2a. Leaf blades hairy, especially below, 3-12 mm wide; spikes usually
overlapping; terminal spike gynecandrous or staminate ... *C. gynodynama*
    2b. Leaf blades glabrous, 2.2-4.5(-6) mm wide; spikes usually distant;
terminal spike staminate .... *C. pellita*
1b. Perigynium beak teeth 0.5-3 mm; perigynia 4.8-12 mm long; style persistent
(see illustration p. 30); range E of Cascades or disturbed places in Portland
    3a. Perigynium beak teeth (1.2-)1.5-3 mm long; perigynia sparsely pubescent
....*C. atherodes*
    3b. Perigynium beak teeth (0.4-)0.6-1.7 mm long; perigynia densely
pubescent
        4a. Upper ♂ scales with spreading white hairs and with apex short-awned;
introduced to disturbed areas around Portland .... *C. hirta*
        4b. Upper ♂ scales glabrous or with appressed hairs near tip, and with
apex lacking awns; native E of the Cascades .... *C. sheldonii*

Key C3: Perigynia hairy; leaf sheath fronts glabrous to scabrous; leaf blades glabrous

1a. Perigynium beak teeth (1.2-)1.5-3 mm long ....*C. atherodes*
1b. Perigynium beak teeth < 1 mm long

2a. Perigynia with only a few scattered hairs; plant cespitose to short-rhizomatous .... *C. luzulina*
2b. Perigynia densely hairy; plant distinctly rhizomatous
   3a. Habitat excessively drained soils, usually pumice, in the Cascades; fertile culms phyllopodic .... *C. halliana*
   3b. Habitat diverse wetlands, widespread; fertile culms aphyllopodic
      4a. Leaves 2-4.5 mm wide at mid-length, with midvein forming a distinct keel on the lower surface; habitat diverse wetlands including bogs and fens .... *C. pellita*
      4b. Leaves 0.2-2(-2.2) mm wide at mid-length (measured as naturally folded), with midvein forming an inconspicuous keel; habitat bogs and fens .... *C. lasiocarpa*

Key D: *Kobresia*; perigynia open on one side; short, cespitose plants of high elevations
1a. Inflorescences simple, unbranched; basal leaf sheaths bladeless ....
....*K. myosuroides*
1b. Inflorescences branched; basal leaf sheaths with remains of dead blades attached .... *K. simpliciuscula*

Key E: Styles persistent; perigynia usually ± inflated
1a. Perigynia corky thickened; introduced to disturbed sandy soils in Portland ..
.... *C. pumila*, p. 411
1b. Perigynia not corky-thickened; range and habitat various
   2a. Perigynia obscurely veined, veins if present not running onto the beak; perigynium beaks 0.3-0.8 mm long .... *C. saxatilis*
   2b. Perigynia strongly veined; perigynium beaks (0.5-)1-4.5 mm long
      3a. Lower lateral spikes nodding on drooping stalks
         4a. Perigynium beak teeth spreading or strongly curved, 1.3–2.8 mm long; stem base brown .... *C. comosa*
         4b. Perigynium beak teeth erect, straight, 0.3–0.9 mm long; stem base red-purple.. *C. hystericina*
      3b. Lower lateral spikes generally erect, stalks sometimes 0
         5a. Spikes crowded, overlapping, except sometimes the lowest ....
....*C. retrorsa*
         5b. Spikes well separated
            6a. Leaves hairy at least at top of leaf sheath front ..... *C. atherodes*
            6b. Leaves glabrous
               7a. Perigynia not inflated; plants cespitose .... *C. distans*
               7b. Perigynia inflated; plants cespitose or rhizomatous
                  8a. Perigynia gradually tapered to beaks 0.5-1.6 mm long; leaves (5.5-)8.5-21 mm wide, glaucous; ligules much longer than wide; NE WA .... *C. lacustris*
                  8b. Plants not as above in all ways; perigynia gradually or relatively abruptly tapered to beaks (1-)1.1-3 mm long;

leaves (1-)1.2-12(-15) mm wide, green or glaucous; ligules
as long as wide or longer than wide; widespread

9a. Dorsal surfaces of basal leaf sheaths with many,
regularly spaced crosswalls between the veins, giving
the appearance of brickwork; most mature perigynia
spreading at right angles from the spike axis

    10a. Foliage green; upper surface of leaf blades smooth
to scabrous, not papillose at 20X; widespread and
abundant in the PNW .... *C. utriculata*

    10b. Foliage strongly glaucous; upper surface of leaf
blades densely papillose at 20X; rare and local in
montane bogs and wet meadows of NE WA ....
.... *C. rostrata*

9b. Dorsal surfaces of basal leaf sheaths with few,
irregularly spaced crosswalls between the veins; mature
perigynia ascending to somewhat spreading but not at
right angles to spike axis

    11a. Longer perigynia 4.0-7.5(-8.2) mm long; perigynia
contracted to the beak; range mainly E of crest of
Cascades .... *C. vesicaria*
Note: plants in the Cascades are often intermediate
between this and the following species)

    11b. Longer perigynia 7.5-10.1 mm long; perigynia
tapering gradually to the beak; range mainly W of
crest of Cascades...................................... *C. exsiccata*

Key F: Stigmas 3; perigynia not pubescent; styles deciduous
Although the length of the sheath of the lowest inflorescence bract is usually a
diagnostic trait, it is occasionally inconsistent. If your plant can't be keyed to a
plausible species, try going the other way at lead 5.

1a. Lateral spikes each with a single perigynium ....Key A5, p. 40

1b. Lateral spikes each with 2+ perigynia

    2a. Leaf blades hairy .... Key F1, p. 46

    2b. Leaf blades glabrous

        3a. Perigynia 9-15 mm long; inflorescences consisting of a single, dense,
much-branched head with many spikes; female inflorescences at least 2
cm wide; habitat sandy soil, usually near coast .... Key F2, p. 46

        3b. Perigynia to 8 mm long; inflorescences consisting of separate spikes
not crowded into a single head, < 1.5 cm wide; habitats various

            4a. Lowest ♀ scales leaf-like and green, similar to the leaf-like lowest
inflorescence bract, 10+ mm long .... *C. cordillerana*

            4b. Lowest ♀ scales not leaf-like, < 10 mm long

                5a. Lowest inflorescence bract with a +/- loose sheath > 4 mm long.
.... Key F3, p. 46

                5b. Lowest inflorescence bract lacking a sheath or with a sheath < 4
mm long

6a. Leaves 8-23 mm wide

    7a. Perigynia strongly flattened .... *C. mertensii*

    7b. Perigynia ± round or trigonous in cross section, not flat.....

    .... *C. amplifolia*

6b. Leaves 1-8 mm wide

    8a. Perigynia with conspicuous beak 0.8-2.7 mm long.............

    .... Key F4, p. 48

    8b. Perigynia beakless or with a short beak up to 0.6 mm long

        9a. Terminal spike ♂, androgynous, or with male and

        female flowers mixed .... Key F5, p. 49

        9b. Terminal spike gynecandrous or ♀ .... Key F6, p. 50

Key F1: Stigmas 3; leaf blades hairy

1a. Perigynia beakless or nearly so, rounded at tip

    2a. Terminal spike staminate or if gynecandrous then less than 1/4 of the flowers pistillate .... *C. pallescens*

    2b. Terminal spike gynecandrous with at least 1/2 of the flowers pistillate ....

    .... *C. hirsutella*

1b. Perigynia with a distinct beak

    3a. Leaf blades hairy throughout; lowest inflorescence bract with sheath < 5 mm long .... *C. whitneyi*

    3b Leaf blades mostly glabrous, sparsely hairy at base or along margins; lowest inflorescence bract with sheath > 5 mm long ..... *C. mendocinensis*

Key F2: Stigmas 3; perigynia and inflorescences very large; sandy habitats; section *Macrocephalae*

1a. Culms sharply angled, serrate on at least one angle; perigynia with 7-9 ventral nerves; perigynium beak at least half as long as perigynium, with sharp teeth; anthers 2.5-5 mm; native at the coast .... *C. macrocephala*

1b. Culms bluntly angled, smooth; perigynium with ± 12 ventral nerves; perigynium beak 1/3 or less as long as perigynium; anthers 4-6.5 mm; introduced, along rivers from Portland to the coast .... *C. kobomugi*

Key F3: Stigmas 3; sheath of lowest inflorescence bract > 4 mm long, usually +/- inflated

1a. Leaves broad, 6-18 mm wide

    2a. Spikes 3-20+ cm long, drooping; plants 90-200+ cm tall .... *C. pendula*

    2b. Spikes 1-3(-4) cm long, erect; plants to 60 cm tall

        3a. Terminal spike 3-8 mm wide; perigynium beaks straight; perigynia green with reddish or purple spots, or all purplish; habitat moist montane meadows and bogs .... *C. luzulina*

        3b. Terminal spike 2-2.7 mm wide; perigynium beaks usually curved toward the dorsal side; perigynia green; habitat lowland to mid-montane mesic forests .... *C. hendersonii*

1b. Leaves narrower, mostly (0.2-)1.5-7 mm wide

4a. One or more of the inflorescence bracts widely spreading, elongate and conspicuously longer than the inflorescence (usually more than twice as long); spikes crowded, stiffly ascending; perigynia widely spreading or reflexed, distinctly beaked; plants densely cespitose ....
....Key F4, lead 5, p. 49

4b. Inflorescence bracts not widely spreading, usually not conspicuously longer than the inflorescence (sometimes up to twice as long); other characters various

  5a. Leaves 0.2-1 mm wide

    6a. ♀ scales awnless, hyaline or green; ♀ spikes with 1-6 perigynia; native plants along rivers in NE WA .... *C. eburnea*

    6b. ♀ scales awned, brown; ♀ spikes mostly with 10-20+ perigynia; ornamental plants occasionally escaped from cultivation ....
.... *C. comans*, p. 410

  5b. Leaves mostly 1.5-7 mm wide

    7a. Perigynium beaks approximately as long as perigynium bodies; introduced in Seattle .... *C. sylvatica*

    7b. Perigynium beaks shorter than perigynium bodies; range various, widespread

      8a. Perigynia strongly flattened, ovate, AND much larger than the achene; terminal spike gynecandrous

        9a. Plants montane; spikes dangling on delicate peduncles ....
....*C. mertensii*

        9b. Plants alpine or subalpine; spikes ascending to erect, though the entire culm may nod as the perigynia mature ....
.... *C. heteroneura*

      8b. Perigynia ± trigonous, not much larger than the achene; terminal spike ♂ or androgynous

        10a. Plants densely to loosely cespitose

          11a. Perigynia blunt and essentially beakless, 2.3-3 mm long .
....*C. pallescens*

          11b. Perigynia beaked, though the beak may be short if perigynia are small

            12a. Terminal spike 4-10 mm long, 0.7-1.4 mm wide

              13a. Culms 10-60 cm tall; perigynia 2.2-3.5 mm long; plants of montane to subalpine bogs; Wallowas and N WA .... *C. capillaris*

              13b. Culms 5-15 cm tall; perigynia 1.5-2.1 mm long; plants of alpine wetlands; Steens Mt. ..... *C. tiogana*

            12b. Terminal spike 11-66 mm long, 1.2-8 mm wide

              14a. Lateral spikes (4-)5-7(-8) mm wide ..... *C. luzulina*

              14b. Lateral spikes 1.5-4 mm wide

                15a. Largest lateral spikes > (2.5-)3.0 cm long; range SW OR .... *C. mendocinensis*

                15b. Largest lateral spikes 1-2.5(-3) cm long; introduced in Morrow and Umatilla cos., OR ....
.... *C. distans*

    10b. Plants rhizomatous
       16a. Perigynium beak 0.5-1.5 mm long
          17a. Perigynia gradually tapered to the beak; leaves green or yellow-green; fertile shoots phyllopodic; plants short-rhizomatous .... *C. luzulina*
          17b. Perigynia abruptly contracted to the beak; leaves glaucous; fertile shoots aphyllopodic; plants long-rhizomatous .... *C. californica*
       16b. Perigynium beak 0-0.4 mm long
          18a. Foliage green; spike with 15-50 perigynia; perigynia often with reddish speckles; lower spikes arising below the middle of the culm; substrate limestone .... *C. crawei*
          18b. Foliage glaucous; spike with 5-15 perigynia; lacking reddish speckles but sometimes brown-blotched distally; lower spikes arising only well above the middle of the culm (or rarely with one basal spike); substrate not limestone
          19a. Perigynia spindle-shaped, widest in the middle and tapering to both ends, 3-5 mm long; beak straight; habitat montane and coastal bogs, substrate not serpentine .... *C. livida*
          19b. Perigynia ovate or obovate, widest above the middle, 1.9-3.6 mm long; beak often curved toward the back; habitat various, meadows and wetlands, including bogs, substrate often serpentine
          20a. Half or more of filled perigynia with 2 stigmas (filled perigynia have well-developed, hard achenes) .... *C. hassei*
          20b. 90% or more of filled perigynia with 3 stigmas (unfilled perigynia may have 2 stigmas) .... *C. klamathensis*

Key F4: Stigmas 3; perigynia with a conspicuous beak 0.8-2.7 mm long

1a. Perigynia thick-walled and corky; plants strongly rhizomatous; habitat disturbed sandy soil, introduced in Portland .... *C. pumila*, p. 411
1b. Perigynia not thick-walled; plants cespitose to rhizomatous; habitat various including coastal sandy wetlands
    2a. ♀ scales with awns 1-3+ mm long; perigynia ascending .. *C. macrochaeta*
    2b. ♀ scales obtuse to acuminate or with a tiny awn much < 1 mm long; at least lower mature perigynia spreading to reflexed
       3a. Perigynia veinless or obscurely veined, not exceeding the achenes; styles persistent; habitat bogs .... *C. saxatilis*
       3b Perigynia distinctly veined; styles deciduous; habitat various
          4a. ♀ scales brown with pale midrib; terminal spike often gynecandrous, sometimes ♂ .... *C. serratodens*

4b. ♀ scales pale; terminal spike ♂ or androgynous
  5a. Perigynia mostly 2.2-3.3 mm long, spreading, straight or
    nearly so; leaves narrow and channeled, mostly to 3 mm wide;
    widespread but local including N WA ..... *C. viridula* ssp. *viridula*
  5b. Perigynia mostly 3.7-6.2 mm long, most of them strongly
    recurved; leaves wider and flat, mostly 2.5-5 mm wide; rare and
    local in N WA .... *C. flava*

**Key F5: Stigmas 3; terminal spike ♂, androgynous, or with male and
female flowers mixed**
1a. Terminal spike 1.5-3.5 mm wide, ♂; perigynia ± round to trigonous in cross
  section
  2a. At least the upper lateral spikes erect to ascending, sessile or on short
    peduncles; young roots lacking dense yellow hairs
    3a. Scales blackish or dark purplish
      4a. Perigynia flat and much longer and wider than the achenes ....
        ....*C. paysonis*
      4b. Periygia not flat, little longer or wider than the achenes
        5a. Styles persistent; perigynia dark brown or blackish at least
          above (rarely green), 2.2-5.5 mm; plants rhizomatous ....
          .... *C. saxatilis*
        5b. Styles deciduous; perigynia green, gray, yellow-brown, or
          orange-brown, 1.9-4.4 mm; plants densely to loosely cespitose
          6a. Perigynia (3-)3.3-4.4 mm long, yellow-brown, orange-brown,
            or gray; habitat dry .... *C. raynoldsii*
          6b. Perigynia 1.9-3.3 mm long, greenish; habitat bogs ..*C. stylosa*
    3b. Scales white-hyaline to medium brown
      7a. Perigynia beakless or nearly so .... *C. pallescens*
      7b. Perigynia with a beak 0.2-1.3 mm long
        8a. Leaves 0.2-0.9(-1) mm wide; perigynia erect .... *C. eburnea*
        8b. Leaves 1.0-3.1(-4.5) mm wide; perigynia widely spreading ......
          .... *C. viridula* ssp. *viridula*
  2b. All lateral spikes spreading to dangling on long, thin peduncles; young
    roots with dense yellow hairs (section *Limosae*)
    9a. ♀ scales much narrower than the perigynia, 1.2-2 mm wide, with awns
      often greater than 1 mm long, making the spike look shaggy ....
      .... *C. magellanica* ssp. *irrigua*
    9b. ♀ scales wider than the perigynia and wrapped around them, or not
      much narrower, 2-3.5 mm wide, acute to slightly pointed, rarely awned,
      the spikes not shaggy
      10a. ♀ scales blackish, wrapping around perigynia at base; lowest
        inflorescence bract bristle-like, inconspicuous, shorter than the
        subtended peduncle .... *C. pluriflora*
      10b. ♀ scales medium brown or reddish, not wrapping around
        perigynia at base; lowest inflorescence bract usually leaf-like and 2+
        cm long .... *C. limosa*

1b. Terminal spike 3.5-6 mm wide, ♂, androgynous, or with irregularly arranged male and female flowers; perigynia strongly flattened, round, or trigonous in cross section

11a. Achenes nearly filling the perigynia; plants strongly rhizomatous

12a. Terminal spike ♂; lateral spikes nodding on long peduncles, ♀ .... *C. magellanica* ssp. *irrigua*

12b. Terminal spike ♂, ♀, or with male and female flowers mixed; lateral spikes sessile and ascending, ♂, ♀, or with male and female flowers mixed .... *C. idahoa*

11b. Achenes much smaller than the perigynia; plants cespitose or short-rhizomatous, forming clumps

13a. Longest ♀ scale awns 1.6-3+ mm long .... *C. macrochaeta*

13b. Longest ♀ scale awns < 1.5 mm long

14a. Perigynia ovate, well over half as wide as long

15a. Ribs of perigynia at the margins .... *C. heteroneura*

15b. Ribs of perigynia displaced from the margins

16a. Spikes 2-4, erect; habitat subalpine and alpine meadows; Wallowas and Elkhorns, NE OR .... *C. paysonis*

16b. Spikes 4-6(-9), drooping gracefully; habitat lowland to montane meadows and forest edges; widespread...*C. mertensii*

14b. Perigynia elliptical, to about half as wide as long

17a. Terminal spike ♂; inflorescence looser, the lower spikes usually stalked and somewhat drooping .... *C. spectabilis*

17b. Terminal spike gynecandrous; inflorescence dense, the lower spikes usually short-stalked and ascending, although the entire culm may nod .... *C. heteroneura*

Key F6: Stigmas 3; terminal spike gynecandrous or ♀; section *Racemosae*, mostly

1a. Perigynia tapering to a beak about 0.5-1 mm long, the beak bidentate, the teeth 0.2-0.4 mm long, minutely bristly; range SW OR .... *C. serratodens*

1b. Perigynia beakless, minutely beaked, or abruptly beaked, the beak 0.2-0.5 mm long; beak teeth absent or short; range various

2a. Plants strongly rhizomatous, often forming large stands or turf

3a. Leaf sheath fronts ladder-fibrillose; perigynia whitish or pale greenish and strongly papillose .... *C. buxbaumii*

3b. Leaf sheath fronts hyaline, not breaking into fibers; perigynia yellowish to brown or blackish, smooth or minutely papillose

4a. Lateral spikes ± erect to ascending; peduncles approximately 1.5 mm long .... *C. idahoa*

4b. Lateral spikes dangling on long, slender peduncles much >1.5 mm long; (section *Limosae*) .... *C. pluriflora*

2b. Plants cespitose to short-rhizomatous, forming clumps

5a. Perigynia dark gold with darker beaks .... *C. atrosquama*

5b. Perigynia brown, dark purplish, or green and yellow
  6a. Achene nearly filling the perigynium; culms 10-40 cm tall
    7a. Inflorescence typically all dark except for whitish hyaline margins of ♀ scales; perigynia dark brown .... *C. albonigra*
    7b. Inflorescence with contrasting light and dark; perigynia green or straw-colored except for dark beak .... *C. media*
  6b. Achene filling half or less of perigynium; culms 25-100 cm tall
    8a. Habitat montane but not subalpine, to 5500 feet elevation; spikes usually 1-4 cm long, all drooping .... *C. mertensii*
    8b. Habitat subalpine to alpine, 5700-9000+ feet elevation; spikes 0.7-2.7 cm long, erect to spreading (though the culm itself may nod)
      9a. Perigynia dark reddish brown, shiny; spikes 0.7-1.0 cm long, sessile to short-pedunculate, crowded .... *C. pelocarpa*
      9b. Perigynia yellow-green to dark brown, dull to shiny; spikes 0.7-2.7 cm long, pedunculate, less crowded ... *C. heteroneura*

Table G. Stigmas 2; achenes lenticular; spikes usually > 1.5 cm. Use this conspectus table to decide if you should use Key G1 or G2 to identify your plant. One unambiguous matching character is enough to indicate which key should be used.

Trait	Go to Key G1, p. 52	Go to Key G2, p. 52
Culm height	up to 0.5 meter	up to 2. 2 meters
Habit	rhizomatous	rhizomatous or densely cespitose
Rhizome diameter	~ 1 mm	1.5-6 mm
Plant base color	pale or dull medium brown	pale, dull medium brown, reddish, chestnut, or blackish
Leaf sheath fronts	hyaline	hyaline or ladder-fibrillose
Leaf sheath fronts, color	white or transparent	white, coppery, or densely red-dotted
Lateral spikes, longest length	1.8(-2.3) cm	(1-)2-15 cm
Sheath of lowest inflorescence bract	(2-)4+ mm long, ± inflated	0-2(rarely to ~5) mm long, not inflated
Staminate spikes	0-1	1 – several
Perigynium texture (when fresh)	succulent (at least at base) or dry	dry
Perigynium apex	angled toward the back or straight	straight (unless bent during collection)
Perigynia per 5 mm on spike	(3-)4-10(-14); not or less crowded	(12-)18-25; crowded

Key G1: Stigmas 2; section *Bicolores*. Two species seem to exist in the PNW, but the taxonomy is unsettled. Immature plants and even some mature ones cannot be identified to species. Such plants can be reported as *Carex* section *Bicolores.*

1a. Terminal spike 0.9 – 2 mm wide; proximal ♂ scales 2 – 3.5 (-4) mm long; mature perigynia succulent throughout, divergent, yellow or orange in life (dark brown and waxy-looking when dry); ♀ scales divergent in mature fruit .... *C. aurea*

1b. Terminal spike (1.8-) 2 – 3.5 mm wide; proximal ♂ scales 3 – 6 (-15) mm long; mature perigynia dry or succulent only at base, ± ascending, green, tan, or whitish, rarely yellowish (similar in color or faded when dry); ♀ scales ± ascending in mature fruit .... *C. hassei*

Key G2: Stigmas 2; section *Phacocystis*, mostly.

1a. Perigynia very thick-walled and hard, resisting puncture, brown or yellow-brown; achene often with an indentation like a dented beer can on one or both margins; range mainly W of the Cascades

    2a. Spikes shorter (the lowest 1.8-5 cm long), brown, blunt at the base, straight but dangling on a peduncle 2-8 cm long; leaf sheath fronts hyaline; leaves dying in winter; habitat coastal salt marshes, also fresh water along lower Columbia River .... *C. lyngbyei*

    2b. Spikes longer (the lowest 2.5-15 cm long), dark brown to blackish, tapering at the base, drooping and sessile or on a straight peduncle to 3 cm (rarely to 9 cm) long; leaf sheath fronts ladder-fibrillose; leaves evergreen; habitat fresh or brackish water ....*C. obnupta*

1b. Perigynia membranous or papery to somewhat tough, generally green or tawny, with or without red or purple pigments; achene not indented; range various

    3a. Perigynia veinless or very faintly veined on both faces (but with 2 marginal ribs) ....Key G3, p. 54

    3b. Perigynia distinctly veined on at least one face (in addition to the 2 ribs)

        4a. Plants rhizomatous

            5a. ♀ scales with a prominent, scabrous awn (at least in the lower part of the spike); perigynia often somewhat tough

                6a. Leaf sheaths densely red-dotted and coppery-colored, usually ladder-fibrillose; lowest inflorescence bract much longer than the inflorescence; range SW OR, CA .... *C. barbarae*

                6b. Leaf sheaths usually white-hyaline (occasionally sparsely red-dotted, sometimes coppery), not ladder-fibrillose; lowest inflorescence bract subequal to the inflorescence; widespread E of Cascades, occasionally introduced in W Cascades ....
....*C. nebrascensis*

            5b. ♀ scales lacking a scabrous awn, acute, acuminate, or with a short point; perigynium traits various

7a. Perigynia (1-)1.5-2.1 mm long, green, widely spaced at base of lowest spike; habitat sandy soil at margins of fast-flowing rivers and streams in and west of the Cascades .... *C. interrupta*

7b. Perigynia 2.2-5.5 mm long, brownish, straw-colored, or green, not widely spaced at base of lowest spike; habitat other substrates (rarely sand) in wet meadows, ditches, and streams

  8a. Leaves 2-4.5 mm wide, minutely papillose on the upper surface; ♀ scales shorter than perigynia; beak 0-0.2 mm long; introduced to SW BC, to be looked for in NW WA .... *C. nigra*, p. 411

  8b. Leaves 3-12 mm wide, smooth and often shiny on the upper surface; ♀ scales shorter or longer than perigynia; beak 0.2-0.6 mm long; E of Cascades, occasionally introduced in W Cascades

    9a. *Mature* perigynia with 5-9 veins, thicker and somewhat tough; perigynium beak minutely bidentate; lower leaf sheaths not ladder-fibrillose .... *C. nebrascensis*

    9b. *Mature* perigynia with 1-3 weak veins, thinner and not tough; perigynium beak not bidentate; lower leaf sheaths usually ladder-fibrillose.................................... *C. angustata*

4b. Plants cespitose

  10a. Plant bases blackish or dark reddish; leaf sheath fronts strongly ladder-fibrillose; perigynia sessile and widest above middle; habitat boulders and cobbles in the scour zone of fast-moving, seasonally flooding rivers, also irrigation ditches .... *C. nudata*

  10b Plant bases greenish or brownish; leaf sheath fronts hyaline; perigynia stipitate and widest at or below middle; habitat silts, sands, or gravels at edges of slow-moving or still water, including drawdown zones of reservoirs .... *C. kelloggii* (with 3 varieties)

    11a. Habitat mostly coastal wetlands and ditches (rarely introduced inland); ♀ spikes 4-6 mm wide, crowded, the lower ones much longer than the internodes between them ..... *C. k.* var. *limnophila*

    11b. Habitat non-coastal, mainly montane to subalpine; pistillate spikes 2-4 mm wide, usually less crowded

      12a. Perigynium bodies green throughout, 5-7 veined on each face; beak purple-brown at very tip only (or occasionally with a little dark brown farther down the tip), stipe > 0.2 mm long .... *C. k.* var. *kelloggii*

      12b. Perigynium bodies spotted purple-brown on apical half, 1-3 veined on dorsal surface; beak entirely purple-brown (or occasionally with some green on the tip), stipe < 0.2 mm long .... *C. k.* var. *impressa*

Key G3: Stigmas 2; perigynia lacking veins, or with very faint veins; section *Phacocystis*, mostly

1a. Plants densely cespitose .... Key G2, lead 10, p. 53
1b. Plants rhizomatous to loosely cespitose
  2a. Perigynia inflated toward the top, loosely enveloping the achene
    3a. Perigynia (3-)3.5-5.5 mm long, usually dark purplish, sometimes greenish, not speckled with red; style persistent; (section *Vesicariae*) ... . *C. saxatilis*
    3b. Perigynia 2.5-2.8 mm long, olive-green, olive-brown, or orangish, with reddish speckles; style deciduous; (section *Phacocystis*)......... *C. aperta*
  2b. Perigynia flattened and more tightly enveloping the achene
    4a. Perigynia (1-)1.5-2.1 mm long, green; widely spaced at base of lowest spike; habitat in sandy soil at margins of fast-flowing rivers and streams in and west of the Cascades .... *C. interrupta*
    4b. Perigynia 2-4 mm long, brownish, often with red or black spots, not widely spaced at base of lowest spike; habitat and range various
      5a. Spikes crowded, with lowest internode 0.3-3.5(-4) cm long and lowest two internodes collectively 0.5-4.2(-4.7) cm long .... *C. scopulorum* (with 2 varieties)
        6a. Basal leaf sheaths bladeless and to 15 cm long; culms 35-90 cm long; perigynia ± acute distally; NE WA ... *C. s.* var. *prionophylla*
        6b. Basal leaf sheaths with blades or if bladeless, then to 5 cm long; culms 11-65 cm long; perigynia rounded distally; widespread .... *C. s.* var. *bracteosa*
      5b. Spikes not crowded, with lowest internode 3-20 cm long and lowest two internodes collectively 4.3-26 cm long...................................... .... *C. aquatilis* (with 2 varieties)
        7a. Spikes erect, the lower ones usually entirely ♀, on peduncles up to 4 cm long; beak of the perigynia tawny, up to 0.2 mm long; mainly E of Cascade crest .... *C. a.* var. *aquatilis*
        7b. Spikes drooping, the lower ones usually androgynous, on peduncles up to 11 cm long; beak of the perigynia generally purplish brown, 0.3-0.4 mm long; mainly W of Cascade crest .... *C. a.* var. *dives*

KEY H: Stigmas 2; perigynia unwinged; spikes androgynous, entirely ♀, or with ♂ flowers mixed irregularly among the ♀ flowers. Two versions of Key H are presented. Key H1 emphasizes perigynium traits. Key H2 (p. 57) emphasizes vegetative traits. No matter which key you choose, you will need both perigynia and shoots to identify your specimens.

Key H1: Emphasizing perigynium traits

1a. Mature perigynia swollen with pithy tissue, usually at the base, sometimes on the ventral surface. (Stick a pin in the base to detect the spongy pith.)

2a. Perigynia long-tapered, so that the distinction between beak and body is unclear ....Key H3, p. 58

2b. Perigynia short-tapered or abruptly narrowed to a distinct beak

3a. Lowest inflorescence node producing 2+ spikes or a branch with 2+ spikes ....Key H4, p. 58

3b. Lowest inflorescence node producing a single spike

4a. Leaf sheath fronts red-dotted

5a. Perigynia 2.3-2.5(-2.9) mm long; native plant of bogs................ ....*C. diandra*

5b. Perigynia 3.5-5.5 mm long; ornamental plant occasionally escaping in NW WA and CA .... *C divulsa*

4b. Leaf sheath fronts white-hyaline, lacking red dots

6a. Perigynium beaks very short, 0.2-0.6 mm long, usually < 1/4 the length of the perigynium body .... *C. simulata*

6b. Perigynium beaks longer, usually 1/3 or more the length of the perigynium body

7a. Perigynia with 5-11 dark veins on each surface; perigynium beaks with smooth margins .... *C. jonesii*

7b. Perigynia veinless or nearly so (or often longitudinally wrinkled in *C. douglasii*), the beak margins serrulate ....Key H5, p. 59

1b. Mature perigynia lacking pithy tissue

8a. Perigynia 1-3 per spike, much longer than subtending ♀ scales, the perigynium walls hyaline and fragile; spikes remote, the distance between them usually longer than the spikes .... *C. disperma*

8b. Perigynia > 3 per spike, shorter to somewhat longer than the subtending scales, the perigynium walls membranous to somewhat tough; spikes overlapping

9a. Plants alpine; inflorescence a tight, globose head barely longer than wide .... *C. vernacula*

9b. Plants of low to high elevations, rarely alpine; inflorescence usually longer than wide

10a. Plants long-stoloniferous and inconspicuously rhizomatous; habitat montane bogs, also introduced to coastal cranberry bogs ....*C. chordorrhiza*

10b. Plants cespitose or rhizomatous but not stoloniferous; habitats various

11a. Inflorescence branched at the lowest node or with 2+ spikes that appear to originate at the lowest node

    12a. Inflorescence loose and interrupted; lowest inflorescence branch somewhat elongated; leaf sheath front coppery or brown, red- or rusty-dotted .... *C. cusickii*

    12b. Inflorescence not interrupted; lowest inflorescence branch short, so that two or more spikes appear to arise at the lowest node; leaf sheath front white-hyaline to dingy brownish .... *C. densa*

11b. Inflorescence not branched, producing only a single spike at each node

    13a. Plants densely to loosely cespitose

        14a. Perigynia hidden by the ♀ scales which are as long as and wider than the perigynia; inflorescence usually angled to one side; range entirely W of the Cascades.....*C. tumulicola*

        14b. Perigynia exposed by the ♀ scales, either because the perigynia are spreading or because the scales are shorter or narrower than the perigynia; range E or W of the Cascades

            15a. Perigynia with 5-11 strong veins on each face ....*C. jonesii*

            15b. Perigynia veinless, but with two marginal ribs

                16a. Perigynia uniformly brown and shiny, bulged so that marginal ribs are displaced to the ventral surface ....*C. vallicola*

                16b. Perigynia green with copper-colored center (maturing light brown with dark brown center), dull or ± shiny, backs not bulged, marginal ribs not displaced to the ventral surface.................. *C. hoodii*

    13b. Plants rhizomatous

        17a. Range W of the Cascades, not on sandy or serpentine substrates; marginal ribs often displaced to the ventral surface; inflorescence axis often bent to one side ....*C. tumulicola*

        17b. Range E of the Cascades or IF W of the Cascades, then on sandy or serpentine substrates; marginal ribs not displaced to the ventral surface; inflorescence axis usually erect

            18a. Perigynium beaks 1.2-2.1 mm long AND inflorescence 6-7 mm wide, usually more than 3 times as long as wide; range Chelan and Kittitas cos., E WA ... *C. siccata*

            18b. Perigynium beaks 0.25-1.9 mm long, IF more than 1.2 mm long, then with the female inflorescences not more than 2.5 times as long as wide; widespread................... ....Key H5, p. 59

Key H2: Emphasizing vegetative traits

1a. Leaf sheath fronts strongly cross-corrugated

  2a. Leaf blades (4-)5-11 mm wide; culm approximately 6 mm wide at mid-length, winged; perigynia 3.6-5.2 mm long .... *C. stipata* var. *stipata*

  2b. Leaf blades 1-7 mm wide; culm approximately 1.5-3 mm wide at mid-length, not or only slightly winged; perigynia 2-4 mm long

    3a. Inflorescence elongate, > 2 times as long as wide, dense or interrupted, gold to light or dark brown; leaf sheath front white-hyaline, dotted with red or pale brown, and/or coppery-tinged .... Key H4, p. 58

    3b. Inflorescence ovoid, 1.5-2 times as long as wide, dense, greenish to dark brown; leaf sheath front white-hyaline .... *C. neurophora*

1b. Leaf sheath fronts smooth, rarely very weakly cross-corrugated

  4a. Leaf sheath fronts minutely dotted red, brown, or yellow; plants cespitose .... Key H4, p. 58

  4b. Leaf sheath fronts white-hyaline to green; plants cespitose or rhizomatous

    5a. Plants long-stoloniferous and inconspicuously rhizomatous; habitat montane bogs, also introduced to coastal cranberry bogs .... *C. chordorrhiza*

    5b. Plants cespitose or rhizomatous but not stoloniferous; habitat various

      6a. Habitat alpine; inflorescence dark brown, dense, and globose, the individual spikes not distinguishable .... *C. vernacula*

      6b. Habitat low elevation to subalpine; inflorescence dark to pale, the individual spikes readily distinguishable to somewhat obscure

        7a. Plants densely to loosely cespitose

          8a. Habitat wetlands, marshes, wet meadows; perigynia lance-triangular, long-tapered to a poorly defined beak .... Key H3, p. 58

          8b. Habitat mesic to dry grasslands; perigynia short-tapered or abruptly narrowed to a distinct beak

            9a. Perigynia bulging dorsally; marginal ribs displaced onto the ventral surface; range E of Cascades .... *C. vallicola*

            9b. Perigynia not or slightly bulging; ribs not displaced onto the ventral surface (except sometimes in *C. tumulicola*); range E or W of Cascades

              10a. Inflorescence a dense, ovoid head, the spikes not easily distinguished; perigynia green with copper-colored center (maturing light brown with dark brown center) ... . *C. hoodii*

              10b. Inflorescence elongate, the lower spikes overlapping but easily distinguished; perigynia brown, sometimes with narrow green margins, not darker over the achene . ... *C. tumulicola*

        7b. Plants rhizomatous

          11a. Perigynia with wings 0.1 mm wide or wider; rhizomes with loose, pithy cortex, easily detached when dry; range Chelan and Kittitas cos., E WA .... *C. siccata*

11b. Perigynia without wings, or with flat margins < 0.1 mm
wide; rhizomes with tight cortex, not detaching when dry;
widespread

12a. Plants soft, delicate, with rhizomes 1 mm in diameter,
dull gray-brown or medium brown; habitat usually shaded;
inflorescence linear and much interrupted, the spikes well
separated from each other, usually 5 mm long or less with
1-3 perigynia; perigynium beak to 0.25 mm long ....
....*C. disperma*

12b. Plants coarser, tougher, with rhizomes > 1 mm in
diameter, pale brown to blackish; habitat usually sunny;
inflorescence dense, though sometimes elongated, the
spikes ± closely aggregated, often > 5 mm long, usually
with > 3 flowers; perigynium beak > 0.25 mm long ....
....Key H5, p. 59

Key H3: Stigmas 2; spikes androgynous; perigynia lance-triangular; section *Vulpinae*

1a. Inflorescence > 3 cm long; widest leaves 5-10 mm wide ....
....*C. stipata* var. *stipata*
1b. Inflorescence < 2 cm long; widest leaves up to 5 mm wide
2a. Leaf sheath fronts with a thick, white rim at mouth, not cross-corrugated;
perigynia 3-4.5 mm; SW OR and N CA .... *C. nervina*
2b. Leaf sheath fronts lacking thick rim at mouth, sometimes cross-
corrugated; perigynia 2.5-4 mm long; widespread
3a. Leaves crowded near base of culm; leaf sheath front usually not cross-
corrugated, mouth usually lacking tongue-like extension; margins of
perigynium beaks and upper bodies entire (or very nearly so) ....
.... *C. jonesii*
3b. Leaves generally not clustered near culm base; leaf sheath front
often cross-corrugated, mouth with tongue-like extension; margins of
perigynium beaks and upper bodies usually serrate on at least one side.
....*C. neurophora*

Key H4: Stigmas 2; spikes androgynous; inflorescence branched at the lowest node, producing a side branch that may be so short that 2+ spikes appear to originate at the node; leaf sheath fronts white-hyaline, dotted red, brown, or yellow, and sometimes also cross-corrugated; plants cespitose

1a. Leaf sheath fronts coppery-tinged toward the mouth, also red-dotted; lowest
inflorescence node usually with a distinct branch with 5-12 spikes ....
.... *C. cusickii*
1b. Leaf sheath fronts white-hyaline or red-dotted, not coppery-tinged;
inflorescence usually a dense head, if distinctly branched then lowest branch
with up to 5 spikes
2a. Leaf sheath fronts cross-corrugated, white-hyaline

58

3a. Inflorescence usually interrupted; perigynia 2-3.2 mm long; range mainly E of the Cascades .... *C. vulpinoidea*

3b. Inflorescence dense, not interrupted; perigynia 2.8-4 mm long; range entirely W of the Cascade crest .... *C. densa*

2b. Leaf sheath fronts not cross-corrugated, either white-hyaline or red-dotted

4a. Perigynia 2.3-2.5(-2.9) mm long; native plant of bogs .... *C. diandra*

4b. Perigynia 3.5-5.5 mm long; introduced ornamental plant of uplands, occasionally escaping .... *C. divulsa*

Key H5: Stigmas 2; spikes androgynous, ♀, or mixed; plants rhizomatous, plants mostly dioecious; section *Divisae*

1a. Perigynia short and squat, 1.8-2.8 mm long, shiny, dark, the beak generally < 1/4 the length of the perigynium body; habitat wetlands with soil submerged or moist at rhizome depth all year long ....*C. simulata*

1b. Perigynia longer, 2.6-4.2 mm long, shiny or dull, the beak > 1/3 the length of the perigynium body; habitat uplands to seasonal wetlands that dry out at rhizome depth in summer

2a. Rhizomes < 2.1 mm thick, brown; leaf blades flat with involute tips, or involute throughout

3a. Perigynia 3.5-4.6 mm long, beaks 0.9-1.8 mm long, ± equal to the body; stigmas persistent, styles exserted from perigynia; inflorescences on ♀ plants 1.5-3.5 cm long, 1.3-2.7 cm wide; widespread E of Cascades .... *C. douglasii*

3b. Perigynia 2.5-3.5 mm long, beaks 0.4-1(-1.2) mm long, shorter than body; stigmas deciduous, styles not exserted from perigynia; inflorescences on ♀ or bisexual plants < 2 cm long, < 1 cm wide; rare and local E of Cascades .... *C. duriuscula*

2b. Rhizomes > 2 mm thick, dark brown to blackish; leaf blades flat or V-shaped in cross section

4a. Habitat coastal sands; ♀ scales usually dark and shiny; ♀ inflorescences usually 1-2 cm wide, ovate; perigynia generally shiny; longest anther awns 0.2-0.4 mm .... *C. pansa*

4b. Habitat inland (but within a few miles of the coast in SW Oregon), in ± alkaline or serpentine soils, including sand; ♀ scales usually tan or dull pale brown; ♀ inflorescences usually < 1(-1.5) cm wide, elliptic or elongate; perigynium ± dull; longest anther awns 0.1-0.2 mm ....
.... *C. praegracilis*

Key I: Stigmas 2; spikes gynecandrous; perigynia unwinged
1a. Spikes crowded in a single, dense head, not easily distinguished ....
.... Key I1, p. 62
1b. Spikes remote to overlapping, but individual spikes easily recognized
  2a. Plants rhizomatous
    3a. Beak (or top of perigynium; it may be virtually beakless) curved; perigynium surface often papillose at least near beak; sheath of lowest inflorescence bract (0-)2-5+ mm long .... *C. hassei*
    3b. Beak not curved; perigynium surface smooth; sheath of lowest inflorescence bract 0-1 mm long
      4a. ♀ scales not hyaline, usually blackish to green or brown, usually narrower and/or shorter than the body of the perigynium; spikes ± stalked, 1-5+ cm long .... section *Phacocystis*; Key G2, p. 52
      4b. ♀ scales hyaline, white with green midvein, subequal to or longer than the body of the perigynium; spikes all sessile, crowded together, 0.4-0.9 cm long; Okanagan Co., WA .... *C. tenuiflora*
  2b. Plants cespitose
    5a. Spikes 3+ times as long as wide, usually stalked
      6a. Perigynium beaks 0.9-2.5 mm long, 28-60% of perigynium length .
.... section *Deweyanae*; Key I2, p. 62
      6b. Perigynium beaks 0.1-0.6 mm long, < 20% of perigynium length ..
.... section *Phacocystis*; Key G2, p. 52
    5b. Spikes < 3 times as long as wide, sessile
      7a. Mature perigynia strongly spreading to reflexed
        8a. Perigynium beak short, 0.5-1.2 mm long, beak length 20-44% of perigynium length
          9a. Perigynia +/- gradually tapered to the beak; wider leaves 2-4 mm wide .... *C. arcta*
          9b. Perigynia abruptly narrowed to the beak; wider leaves 1-2.4 (-2.7) mm wide .... *C. interior*
        8b. Perigynium beak long, (0.85-)0.95-2 mm long, beak length (35-)38-60% of perigynium length; perigynium tapering gradually to the beak; widest leaves 1-3.3+ mm wide......................................
.... *C. echinata* (with two subspecies)
          10a. Inflorescences more open, lowest internode longer than lowest spike .... *C. e.* ssp. *echinata*
          10b. Inflorescences dense, lowest internode shorter than lowest spike
            11a. Perigynia 2.9-3.6(-4) mm long, ventral surface usually veinless; widest leaves 1-2.4(-2.7) mm wide; widespread but not coastal..*C. e.* ssp. *echinata*
            11b. Perigynia larger (3.1-)3.5-4.8 mm, ventral surface usually with 2-12 veins; widest leaves (1.7-)2.3-3.3 mm wide; coastal .... *C. e.* ssp. *phyllomanica*
      7b. Mature perigynia appressed to ascending (sometimes seemingly spreading because the beaks are bent back)

12a. Perigynia 3.3-5.3 mm long .... Key I2, p. 62
12b. Perigynia 1.5-3(-3.5) mm long
   13a. Spikes crowded, overlapping, the lowest internode no longer than the lowest spike
      14a. Perigynium beakless or nearly so; range Okanogan Co, WA .... *C. tenuiflora*
      14b. Perigynium with a distinct beak; range various
         15a. Perigynium beaks smooth on margins
            16a. Beak to 0.5 mm .... *C. praeceptorum*
            16b. Beak 1-1.6 mm .... *C. integra*
         15b. Perigynium beaks serrulate on margins
            17a. Inflorescence green to tan; perigynia +/- spreading when ripe, exposed by the ♀ scales, 2-3(-3.5) mm long, 1.2-1.5 mm wide .... *C. arcta*
            17b. Inflorescence brown (with green perigynia when young), perigynia appressed, +/- hidden by ♀ scales, 3.5-4.2 mm long, 1-1.2 mm wide .... *C. leporinella*
   13b. At least lower spikes remote, the lowest internode longer than the lowest spike
      18a. Perigynium beaks 0.4-1.1(-1.3) mm long... *C. laeviculmis*
      18b. Perigynium beaks up to 0.5 mm long
         19a. Dorsal suture darker than the surrounding perigynium surface; ♀ scales light brown with lighter center and hyaline edges .... *C. praeceptorum*
         19b. Dorsal suture about the same color as the surrounding perigynium surface; ♀ scales white-hyaline (often tinged brownish with age) with green center
            20a. Leaves obviously glaucous ....
            .... *C. canescens* ssp. *canescens*
            20b. Leaves green or intermediate in color
               21a. Inflorescence more compact, sometimes lowest spikes separate; dorsal suture inconspicuous, usually shorter than the beak, 0-0.4(-0.7) mm long; perigynia (5-)10-20+ per spike, appressed, the tips not interrupting the outline of the spike ...
               .... *C. canescens* ssp. *canescens*
               21b. Inflorescence more elongate, most spikes well separated; dorsal suture readily visible, usually as long as the beak or extending onto the top of the body, usually 0.4-0.8 mm long; perigynia usually 5-10 per spike, slightly more spreading, the tips interrupting the outline of the spike ....
               .... *C. brunnescens* ssp. *brunnescens*

Key I1: Stigmas 2; spikes gynecandrous; perigynia unwinged; inflorescence a tight head

1a. Perigynia beakless or nearly so, plants loosely cespitose to rhizomatous; Okanogan Co., WA .... *C. tenuiflora*
1b. Perigynia with distinct beak; plants tightly cespitose; range various
    2a. Longest inflorescence bracts > 3 times as long as the inflorescence; perigynium beak 3-5 mm long, longer than the perigynium body ....
    ....*C. sychnocephala*
    2b. Longest inflorescence bracts rarely longer than the inflorescence and never twice as long; perigynium beak 1-1.6 mm long, shorter than the perigynium body
        3a. Inflorescence ovoid to elongate, green to straw-colored or pale brown; spikes 5-15; margins of beaks serrulate .... *C. arcta*
        3b. Inflorescence ± pyramidal, globose, or ovoid, dark brown or a mix of green and brown or black; spikes 3-9; margins of beaks entire
            4a. Inflorescence black and green or black and brown, compact, the base usually +/- truncate; perigynium wings absent .... *C. illota*
            4b. Inflorescence brown or brown and green, elongate, the base usually +/- tapering; perigynium wings very narrow (0.05-0.2 mm wide) ....
            .... *C. integra*

Key I2: Stigmas 2; spikes gynecandrous; perigynia unwinged, 3.3-5.3 mm long; section *Deweyanae*. Beak length should be measured from the achene top to the beak tip. Immature plants and even some mature ones cannot be identified to species. Unidentified specimens can be reported as *Carex* section *Deweyanae*.

1a. Ligules on uppermost leaves of fertile culm 0.9-2.2 mm, ± rounded, about as long as wide; anthers (1.8-)1.9-2.2 mm long; longest inflorescence with 2-5 spikes; mountains of NE WA .... *C. deweyana* var. *deweyana*
1b. Ligules on uppermost leaves of fertile culm (2.1-)3.1-9.1 mm, ± triangular, much longer than wide; anthers 1.4-1.9(-2.2) mm long; longest inflorescence with (4-)5-9 spikes; widespread in the PNW
    2a. Beaks 0.9-1.5(-1.7) mm, 28-38% of perigynium length; bodies of ♀ scales (2.3)2.7-3.8 mm ....*C. leptopoda*
    2b. Beaks (1.4-)1.6-2.7 mm, 38-50% of perigynium length; bodies of ♀ scales 2.1-2.9(-3.1) mm
        3a. Teeth of perigynium beak 0-0.2(-0.4) mm long ....*C. infirminervia*
        3b. Teeth of perigynium beak (0.2-)0.3-1 mm long .... *C. bolanderi*

Key J: Perigynia winged; stigmas 2; mainly section *Ovales*. Read "Using the *Ovales* Key" (p. 34) before using this key.

1a. Lowest 1-3 bracts of most inflorescences elongate, at least the lowest as long as or usually much longer than inflorescences; inflorescences dense, head-like
    .... Key J1, p. 63
1b. Lowest 1-3 bracts of most inflorescences usually inconspicuous, shorter than the inflorescences; inflorescences dense or elongate

2a. Leaf sheath fronts green and veined nearly to the top, often with a short white-hyaline triangle (< 6 mm long) at the top .... Key J2, p. 64
2b. Leaf sheath fronts white-hyaline for at least 10 mm at the top (often transparent if plant is fresh)
   3a. Longer perigynia (6-)6.3-8.5 mm long .... Key J3, p. 64
   3b. Longer perigynia 2-5.4(-6) mm long
      4a. Inflorescence a dense, round to slightly oval head, although occasionally the lowest spike or two may be separated
         5a. Heads dark (brown, dark brown, reddish brown, or blackish) OR with contrast of dark ♀ scales and pale perigynia. This is an arm's-length trait .... Key J4, p. 65
         5b. Heads pale (whitish, green, straw-colored, or tan) without contrasting dark ♀ scales. If you're unsure, it's not pale ,.............
         .... Key J5, p. 67
      4b. Inflorescence ± elongate, varying from a longer oval with overlapping spikes to a line of separated spikes
         6a. Spikes green to straw-colored AND with a fine texture AND ♀ scales acuminate and revealing acuminate perigynium beaks; beaks winged and ciliate-serrulate nearly to the tip; wings usually narrowed or absent below middle of perigynium; range Willamette Valley, OR, Puget Trough, and coastal and NE WA ...
         ....*C. scoparia* var. *scoparia*
         6b. Spikes dark or if pale, then differing in some way from above
            7a. Inflorescence erect .... Key J6, p. 67
            7b. Inflorescence bent to one side or nodding .... Key J7, p. 69

Key J1: Perigynia winged; lower inflorescence bracts usually longer than the inflorescences; inflorescences dense, head-like. Late-season shoots with atypical, elongated inflorescence bracts are not keyed here.
1a. Beak distinctly longer than (sometimes twice as long as) the perigynium body; distance from top of achene to tip of beak 3-5 mm; range E WA and Harney Co., OR ....*C. sychnocephala*
1b. Beak little if at all longer than perigynium body; distance from top of achene to tip of beak 1.1-2.5 mm; range E or W of the Cascades
   2a. Inflorescence angled to one side; lowest inflorescence bract ± leaf-like; range W of the Cascades .... *C. unilateralis*
   2b. Inflorescence erect, located symmetrically at the top of the culm; lowest inflorescence bract usually not leaf-like; E or W of the Cascades
      3a. Beak tip to achene top 1.9-2.5 mm; achenes 0.3-0.4 mm thick; range mainly E of the Cascades, scattered on the W side but not coastal ....
      ....*C. athrostachya*
      3b. Beak tip to achene top 1.2-1.8 mm; achenes 0.5-0.7 mm thick; range coastal, occasionally inland in SW OR
         4a. Perigynia with 3-8 strong veins on the ventral surface, extending beyond the top of the achene; perigynia not leathery; range near the coast in SW OR .... *C. harfordii*

4b. Perigynia with 0-3 veins on the ventral surface, reaching at most the top of the achene; perigynia +/- leathery; range SW OR north along coast to NW WA .... *C. subbracteata*

Key J2: Perigynia winged; leaf sheath fronts green-veined nearly to the top, often with a triangular white-hyaline area extending at most 6 mm below the top of the leaf sheath front.

1a. Perigynia lanceolate AND wing much narrowed or lacking below middle of body
  2a. Widest leaves < 3 (rarely 4) mm wide; spike apex acute to rounded ....*C. scoparia* var. *scoparia*
  2b. Widest leaves 3-7 mm wide; spike apex rounded ....*C. tribuloides* var. *tribuloides*
1b. Perigynia lanceolate to ovate or obovate, wing usually extending to base of perigynium; if wing much narrowed below middle, perigynium elliptic to ovate
  3a. Perigynia obovate, the body widest above middle, narrowed abruptly to the beak; wing 0.5-0.8 mm wide; introduced to coastal wetlands . *C. longii*
  3b. Perigynia lanceolate to ovate, the body widest near middle, narrowed more gradually to the beak; wing 0.2-0.6 mm wide; native; range various, including coast
    4a. ♀ scales acuminate; perigynia 4.2-5.5(-6) mm, elliptic, tapering gradually to the beak .... *C. scoparia* var. *scoparia*
    4b. ♀ scales obtuse to acute; perigynia 3.2-4.2 mm, ovate, tapering more abruptly to the beak .... *C. feta*

Key J3: Perigynia winged; longer perigynia (6-) 6.3-8.5 mm long

1a. Perigynia planoconvex or biconvex, 0.5-0.9 mm thick; wing 0.3-0.5 mm wide; habitat lowlands to subalpine
  2a. Inflorescence a round to oval head, spikes all crowded or sometimes the lowest distinct .... *C. pachycarpa*
  2b. Inflorescence +/- elongate, spikes overlapping but not crowded
    3a. Perigynia rarely as much as 6 mm long, with 0-4(-7) ventral veins, these usually shorter than the achene .... *C. praticola*
    3b. Perigynia usually > 6 mm long, with 4-10 ventral veins, at least 3 of them longer than the achene
      4a. ♀ scales covering more than half the perigynium beak, white-hyaline or with white-hyaline margin 0.2-0.7 mm wide; uncommon but widespread .... *C. petasata*
      4b. ♀ scales covering less than half the perigynium beak, with white-hyaline margin 0-0.2 mm wide; rare and local .... *C. davyi*
1b. Perigynia flat except where distended by the achene, 0.25-0.5 mm thick; wing 0.3-0.9 mm wide; habitat subalpine to alpine
  5a. Perigynia widest at or above middle of total length, contracted to a relatively short beak .... *C. proposita*

5b. Perigynia widest below middle of total length, contracted to a relatively long beak
  6a. Perigynia much longer than wide; distal 1 mm or more of beak unwinged, brown, and parallel sided .... *C. haydeniana*
  6b. Perigynia little longer than wide; distal 0-0.7 mm of beak unwinged, brown, and parallel-sided .... *C. straminiformis*

Key J4: Perigynia winged; inflorescence a dense, dark head; section *Ovales*, mostly

1a. Perigynium wings 0.4-1 mm wide; perigynia 1.8-3.5 mm wide
  2a. Perigynia planoconvex, 0.5-0.8 mm thick, wings widest on the beak and much narrowed below middle .... *C. pachycarpa*
  2b Perigynia flat except where distended by the achene, 0.2-0.5(-0.6) mm thick, wings broad to near the base
    3a. Perigynia widest near the middle of total length .... *C. proposita*
    3b. Perigynia widest below the middle of total length
      4a. Beak usually winged and ciliate-serrulate almost to the tip, sometimes unwinged, brown, and parallel-sided for the distal 0.5-0.7 mm; inflorescence with spikes easily distinguished, green or brown, often with contrast of dark ♀ scales and paler perigynia ....
.... *C. straminiformis*
      4b. Beak unwinged, brown, and parallel-sided for the distal 1+ mm, entire for 0.3-0.6 mm; inflorescence with spikes crowded and hard to distinguish, dark or contrasting green and dark
        5a. Perigynia 4-6.5 mm long, usually at least (2.3-)2.6 mm from beak tip to achene; at or above timberline .... *C. haydeniana*
        5b. Perigynia 3-4.5(-5.2) mm long, usually no more than 2.5 mm from beak tip to achene; montane to near timberline ....
.... *C. microptera*
1b. Perigynium wings 0.1-0.5 mm wide; perigynia 1.1-2.4 mm wide
  6a. Spikes androgynous; perigynium bodies copper-colored with green margins (maturing dark brown with light brown margins); section *Phaestoglochin* .... *C. hoodii*
  6b. Spikes gynecandrous; perigynia usually ± uniformly colored or with body paler than wings; section *Ovales*
    7a. Perigynium beaks winged to the tip, or with very short (<0.4 mm long) unwinged, brown, parallel-sided tip
      8a. Perigynia (3.8-)4.4-6 mm long; achenes 1.7-2.4 mm long ....
.... *C. pachycarpa*
      8b. Perigynia 2.4-4.3(-4.6) mm long; achenes 1-2 mm long
        9a. Lowest two internodes of inflorescence collectively < 1/3 of total inflorescence length; habitat moist to wet sites near the coast at 0-3000 feet elevation .... *C. harfordii*
        9b. Lowest two internodes of inflorescence collectively > 1/3 of total inflorescence length; habitat mesic to dry sites in mountains, 1200-9000 feet elevation .... *C. preslii*

7b. Perigynium beaks with longer (>0.4 mm long) unwinged, brown, and parallel-sided tip

10a. Perigynia with (8-)10-20 veins on the dorsal surface, the veins often sunken; perigynia planoconvex, 0.5-0.8 mm thick, cream-colored to light brown, often with green wings; wings (0.2-)0.3-0.4 (-0.5) mm wide but often narrow at and below middle of perigynium body; habitat at mid to high elevations in mountains .. *C. pachycarpa*

10b. Perigynia with 0-9(-13) veins on the dorsal surface; veins, if present, raised; other traits various

11a. Perigynia with 3-8 strong veins on the ventral surface, extending above the top of the achene

12a. Habitat at low elevations (0-3,000 feet) mainly near the coast; perigynia (0.5-)0.6-0.7 mm thick .... *C. harfordii*

12b. Habitat montane to alpine (4000-9,000+ feet); perigynia 0.3-0.5(-0.6) mm thick

13a. Distance from achene top to perigynium beak tip (1.5-)2-2.3 mm; habitat montane to subalpine .... *C. abrupta*

13b. Distance from achene top to perigynium beak tip (2.3-) 2.6-3.8 mm; habitat subalpine to alpine ...... *C. haydeniana*

11b. Perigynia with 0-8 veins on the ventral surface, the veins if present faint and/or not more than 2 of them extending above the top of the achene

14a. Perigynia 4-6.5 mm long, distance from beak tip to top of achene (2.3-)2.6-3.8 mm; habitat subalpine to alpine .... *C. haydeniana*

14b. Perigynia 2.7-4.7(-5.7) mm long; distance from beak tip to top of achene 1-2.5(-2.8) mm; habitat lowland to subalpine

15a. Perigynium wing entire, 0-0.2 mm wide .... *C. integra*

15b. Perigynium wing minutely ciliate-serrulate, 0.2-0.5 mm wide

16a. Perigynia many, crowded, appressed, and flat except over the relatively small achene, 0.3-0.5 mm thick, thus making the inflorescence more fine-textured; spikes usually green and black .... *C. microptera*

16b. Perigynia fewer, less crowded, appressed to spreading, planoconvex and +/- filled by the relatively large achene, (0.4-)0.5-0.7 mm thick, thus making the inflorescence coarse-textured; spikes usually coppery, brown, or green and black

17a. Perigynium beaks of 2 kinds in the same inflorescence, some +/- flat and winged nearly to the tip, others with tip cylindric, unwinged, smooth, brown, and parallel-sided for >0.4 mm; habitat montane to subalpine, +/- dry .... *C. preslii*

17b. All perigynium beaks with tips cylindric, unwinged, smooth, brown, and parallel-sided >0.4 mm; habitat at low to mid elevations, moist to mesic

18a. Perigynia spreading; spikes appearing star-shaped from above; ♀ scales 2.2-3.4(-4.2) mm long; perigynia not leathery; range widespread ....*C. pachystachya*

18b. Perigynia +/- ascending, spikes not appearing star-shaped from above; ♀ scales 3.4-4.5(-5.7) mm long; perigynia +/- leathery; range near the coast.. ....*C. subbracteata*

Key J5: Perigynia winged; inflorescence a dense, pale head

1a. Perigynia 3.5-4.5 times as long as wide

2a. Perigynium beaks winged nearly to the tip, perigynia 0.15-0.35 mm thick ....*C. crawfordii*

2b. Perigynium beak tips unwinged, parallel-sided, and entire for at least 0.4 mm; perigynia 0.35-0.45 mm thick .... *C. athrostachya*

1b. Perigynia 1.8-3.5 times as long as wide

3a. Inflorescence relatively fine-textured in appearance; perigynia 0.9-1.9 mm wide, 0.3-0.45 mm thick; habitat lowlands to subalpine

4a. Perigynia tapering to a distinct beak; inflorescence nearly always oval ....*C. subfusca*

4b. Perigynia tapering more gradually to the tip; inflorescence always a dense, roundish head .... *C. athrostachya*

3b. Inflorescence relatively coarse-textured in appearance; perigynia 1.4-3.4 mm wide, 0.3-0.8 mm thick; habitat montane to subalpine

5a. Perigynium wings 0.4-1 mm wide; perigynia flat except where distended by the achene, 0.3-0.5 mm thick .... *C. straminiformis*

5b. Perigynium wings 0.1-0.4(-0.5) mm wide; perigynia planoconvex, 0.5-0.8 mm thick .... *C. pachycarpa*

Key J6: Perigynia winged; inflorescence slightly to very elongate, erect

1a. Lowest inflorescence node producing 2+ spikes; ornamental escaping in NW WA; section *Phaestoglochin* .... *C. divulsa*

1b. Lowest inflorescence node producing a single spike; widespread natives; section *Ovales*

2a. Perigynium wing 0.4-1 mm wide; perigynia 1.8-3.4 mm wide

3a. Plants of low elevations, usually below 2000 feet

4a. Perigynia 2.3-3.2 mm wide, broadly ovate to orbicular .... *C. brevior*

4b. Perigynia to 2.1 mm wide, lanceolate to ovate .... *C. leporina*

3b. Plants of high elevations, usually above 5000 feet

5a. Perigynia widest below middle of entire length .. *C. straminiformis*

5b. Perigynia widest near middle of entire length .... *C. proposita*

2b. Perigynium wing 0.1-0.4(-0.6) mm wide; perigynia 0.9-2.2 mm wide

    6a. Inflorescences silvery or pale green to tan, 3.5-8 cm long; leaf sheath front rounded to acute at summit, prolonged at least 3 mm above attachment point of the leaf blade; plants coarse, usually > 60 cm tall... .... *C. fracta*

    6b. Inflorescences pale to dark, 1-5 cm long; leaf sheath front often U-shaped at summit, usually not prolonged or prolonged < 2.8 mm above attachment point of leaf blade; plants coarse or delicate, usually < 60 cm tall

        7a. Lowest two inflorescence internodes collectively at least 10 mm long

            8a. ♀ scales distinctly shorter than the perigynia; perigynia 2-4(-4.3) mm long .... *C. subfusca*

            8b. ♀ scales about as long as the perigynia; perigynia 3.4-6 mm long

                9a. Habitat subalpine to alpine; perigynia widest near middle of total length

                    10a. Ventral surface of perigynia with 3-8 strong veins that extend to the top of the achene; perigynia opaque, somewhat leathery, (3.7-)4.7-6 mm long .... *C. tahoensis*

                    10b. Ventral surface of perigynia with 0-4 weak veins; perigynia translucent or brown, not especially tough, 3.8-5.2 mm long .... *C. phaeocephala*

                9b. Habitat in lowlands or montane; perigynia widest below middle of total length

                    11a. Tip of perigynium beak white; ventral surface of perigynium usually lacking veins; ♀ scales with white-hyaline margin 0.1-0.3 mm wide .... *C. praticola*

                    11b. Tip of perigynium beak brown; ventral surface of perigynium usually with 3-5+ veins at least as long as the achene; ♀ scales usually lacking white-hyaline margins .... *C. leporina*

        7b. Lowest two inflorescence internodes collectively 3-10 mm long

            12a. Spikes globose; mature perigynia light to dark brown, 2.5-3.8 mm long; beak winged to the tip on all perigynia; achenes 1-1.3 mm long, 0.6-0.9 mm wide .... *C. bebbii*

            12b. Inflorescence differing in some way from above

                13a. Inflorescence fine-textured, usually pale, green, whitish, or light brown; achenes 1-1.6 mm long .... *C. subfusca*

                13b. Inflorescence coarse-textured and darker brown, reddish-brown, blackish, or sometimes greenish; achenes 1.4-2.4 mm long

                    14a. Perigynia 1-1.2 mm wide, boat-shaped; habitat soggy wet in the spring though often drying later .... *C. leporinella*

                    14b. Perigynia 1.3-2.3 mm wide, not boat-shaped; habitat mesic to dry

15a. Ventral surface of perigynia with 3-8 strong veins that extend to the top of the achene; habitat dry subalpine to alpine slopes .... *C. tahoensis*

15b. Ventral surface of perigynia with 0-4 weak veins; habitat mesic lowlands to dry alpine slopes

   16a. Habitat dry alpine slopes; perigynia widest near middle, gold to brown .... *C. phaeocephala*

   16b. Habitat mesic lowlands to dry subalpine meadows; perigynia widest below middle, brown, coppery, or greenish

      17a. Perigynium beaks of 2 kinds in the same inflorescence, some +/- flat and winged nearly to the tip, others with tip cylindric, unwinged, smooth, brown, and parallel-sided for >0.4 mm; habitat montane to subalpine, dry .... *C. preslii*

      17b. All perigynium beaks with tips cylindric, unwinged, smooth, brown, and parallel-sided >0.4 mm; habitat at low to mid elevations, moist to mesic

         18a. Perigynia spreading; spikes appearing star-shaped from above; ♀ scales 2.2-3.4(-4.2) mm; perigynia not leathery; range widespread ....*C. pachystachya*

         18b. Perigynia +/- ascending, spikes not appearing star-shaped from above; ♀ scales 3.4-4.5(-5.7) mm; perigynia +/- leathery; range near the coast .... *C. subbracteata*

Key J7: Perigynia winged; inflorescence elongate, nodding

1a. ♀ scales mostly hyaline, white to pale brown, with green or brown midrib; spikes usually widely separated, 4-10 mm long; rare in N WA .... *C. tenera* var. *tenera*

1b. ♀ scales not mostly hyaline, brown with paler midrib, with or without white margins; spikes usually overlapping, 8.5-20 mm long; widespread

   2a. Tip of perigynium beak white; ventral surface of perigynia usually lacking veins; ♀ scales with white-hyaline margins 0.1-0.3 mm wide.. *C. praticola*

   2b. Tip of perigynium beak brown; ventral surface of perigynia usually with 3-5+ veins at least as long as the achene; ♀ scales usually lacking white-hyaline margins .... *C. leporina*

The counties of Oregon and Washington

Species Accounts

Carex abrupta Mack.

Common name: Abrupt-beak Sedge
Section: *Ovales*   Key: J

KEY FEATURES:
- Cespitose, with gynecandrous spikes & winged perigynia
- Perigynia brown and strongly veined on both faces
- Inflorescence a single dense head
- Meadows at moderate to high elevations in mountains

DESCRIPTION: Habit: Cespitose. **Culms**: 18-70 cm tall. **Leaves**: 1.5-3.7 (-4.9) mm wide. **Inflorescences**: Dense to slightly elongated heads 1.2-2.2 cm long, (6-) 9-20 mm wide. Spikes gynecandrous. ♀ **scales**: Reddish brown with lighter or green midstripe, 2.4-3.9 mm long, shorter than and as wide as or narrower than the perigynia. **Perigynia**: Reddish brown or dark brown, lacking metallic sheen, typically elliptic to lance-ovate, sometimes lanceolate and nearly wingless, usually planoconvex, (2.9-) 3.6-5.4 mm long, 1-2.1 mm wide, with 3+ conspicuous veins that reach or exceed the top of the achene on each face of the perigynium, with wing usually 0.2-0.3 mm wide and sometimes curved toward the ventral surface, making the perigynium boat-shaped. Beak tips coppery or reddish brown, usually unwinged, brown, and parallel-sided for the distal 0.5-0.8 mm; (1.6-) 2-2.3 mm from top of achene to tip of beak. Stigmas 2. **Achenes**: Lenticular, 1.2-1.8 mm long, 0.7-1.1 mm wide, (0.3-) 0.4-0.5 mm thick.

HABITAT AND DISTRIBUTION: Moist meadows and streambanks at moderate to high elevations in mts; Siskiyous, Cascades, Wallowas, and Steens Mt., OR, and north Cascades, WA. OR to ID, S to CA and NV.

IDENTIFICATION TIPS: *Carex abrupta* resembles *C. pachystachya* but has more strongly veined perigynia, especially on the ventral surface, and perigynium wings that may bend in toward the ventral surface more and narrow irregularly toward the beak. *Carex pachystachya* perigynia have a metallic sheen and usually lack veins on the ventral surface; if ventral veins are present, they end at or below top of the achene, or rarely 1 or 2 of them may extend onto the beak. *Carex microptera* has perigynia flat except over the achene, proportionately smaller achenes, and a green and black inflorescence. On Steens Mt., both *C. haydeniana* and *C. microptera* have unusually prominent ventral veins, similar to those of *C. abrupta,* but have either longer or more clearly defined beaks.

COMMENTS: Many SW OR plants that have been identified as *C. abrupta* are odd. They may represent one of the NW CA species such as *C. subbracteata*, but more research is needed.

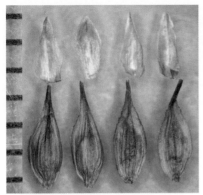

top left: pistillate scales, perigynia
top right: inflorescence
bottom left and right: habit

Carex albonigra Mack.
Common name: Black-and-White Sedge
Section: *Racemosae*   Key: F

KEY FEATURES:
- Inflorescence black or dark brown, the ♀ scales with hyaline margins
- Lower spikes on short peduncles
- Rocky alpine ridges and meadows

DESCRIPTION: Habit: Cespitose. **Culms**: 10-40 cm tall. **Leaves**: 2.5-5 mm wide. **Inflorescences**: 2-4 spikes, each 0.8-2 cm long, the lateral spikes ♀ and erect on short peduncles, the terminal spike gynecandrous. ♀ **scales**: Brown to blackish, about equal to the perigynium, tip acute, the midrib dark like the rest of the scale, margins hyaline, contrasting with the rest of the inflorescence. **Perigynia**: Ovate, more or less flattened except over the achene, dark brown, 3-3.5 mm long, 2-2.5 mm wide, papillose, with a small beak 0.3-0.4 mm long. Stigmas 3. **Achenes**: Trigonous, nearly filling bodies of perigynia.

HABITAT AND DISTRIBUTION: Dry, windblown alpine and subalpine rocky slopes and meadows where little snow accumulates in winter; mts of N WA. AK to NT, S to WA and NM.

IDENTIFICATION TIPS: *Carex albonigra* inflorescences are dark except for pale margins of ♀ scales. Spikes are on stalks so short that they can superficially seem sessile, like the sessile spikes of *C. media* and *C. pelocarpa*. *Carex media* has more contrast between dark scales and greenish or lighter brown perigynia. *Carex pelocarpa* has a shorter, more uniformly dark inflorescence. *Carex atrosquama* has obtuse ♀ scales that are shorter than the perigynia and less conspicuously pale-margined, and its perigynia are a golden brown except at the dark tips.

COMMENTS: Plants previously called *C. albonigra* in the White Mountains and central Sierra Nevada of California have recently been described as a distinct species, *C. orestera* Zika.

top left: perigynia
top right: inflorescence
center left: inflorescence
bottom left and right: habit

Carex amplifolia Boott. in W. J. Hook.
Common name: Big-leaf Sedge
Section: *Anomalae*   Key: F

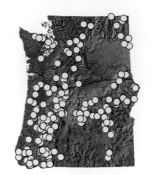

KEY FEATURES:
- Very large plants in rhizomatous patches
- Broad leaves
- Long spikes like *Carex aquatilis* var. *dives,* but perigynia trigonous

DESCRIPTION: Habit: Rhizomatous, forming large populations. **Culms**: 50-100 (-130) cm tall. **Leaves**: (3-) 8-20 mm wide. **Inflorescences**: Lower spikes ♀, ascending to spreading, (1.5-) 3.5-14 cm long, (2.5-) 3.5-6.5 mm wide, upper spikes ♂, 5-9.5 cm long. **♀ scales:** Smaller and narrower than the perigynia, often with a short awn. **Perigynia**: 2.4-3.1 mm long, brownish green, ribbed, obovoid, trigonous or somewhat inflated and more or less round, with narrow beak 0.7-1.1 mm long. Stigmas 3. **Achenes**: Trigonous.

HABITAT AND DISTRIBUTION: Streamsides, road ditches, and other wet areas in conifer woodlands, often in the shade, common W of the Cascades and scattered in mts to the E. S BC to MT, south to CA, also NM.

IDENTIFICATION TIPS: *Carex amplifolia* is a broad-leaved, rhizomatous species forming dense stands in partly shaded wetlands. It has long narrow spikes and 3 stigmas/perigynium. Sterile shoots are very similar to those of *Scirpus microcarpus,* which is often found in the same habitats. Fertile shoots are superficially similar to *C. aquatilis* var. *dives,* which has 2 stigmas/perigynium and short perigynium beaks. *Carex pendula,* invading shaded riparian areas in Seattle and Portland, has similar vegetation but is cespitose and has very long spikes that hang vertically. Its perigynia have short beaks.

COMMENTS: *Carex amplifolia* is a community dominant in some shaded wetland habitats, often with a deciduous overstory component. Rhizomes have been used for basketry. This species can provide an interesting background for the garden, particularly in moist areas, but may be too large and aggressive for small sites. Seeds and plants are available in the horticulture trade.

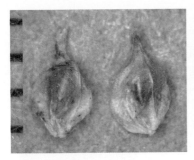

top left: perigynia
top right: inflorescence
bottom: habit

Carex angustata Boott in W. J. Hook.
Common name: Narrow-leaved Sedge
Section: *Phacocystis*   Key: G
Synonyms: *C. eurycarpa* T. Holm

KEY FEATURES:
* Rhizomatous, with purplish black plant bases
* Leaves usually somewhat glaucous, 4-7 mm wide
* Perigynia with 1-3 weak veins on back

DESCRIPTION: Habit: Rhizomatous but the deep rhizomes often not collected. **Culms**: 30-110 cm. Plant bases reddish brown or purplish black. **Leaves**: somewhat glaucous, 4-7 mm wide. Leaf sheath fronts ladder-fibrillose. **Inflorescences**: Lowest inflorescence bract subequal to the inflorescence. Lateral 3-4 spikes ♀, erect, 2.5-7 cm long, 3-5 mm wide. Terminal 1-2 spikes ♂. **♀ scales**: Reddish brown or black, mostly equal to or longer than the perigynia. **Perigynia**: Pale brown with reddish spots, elliptic to obovate, with 1-3 thin but distinct veins at maturity, 2.2-3 mm long with sloping "shoulders," the beak small (0.2-0.5 mm), not bidentate. Stigmas 2. **Achenes**: Lenticular.

HABITAT AND DISTRIBUTION: Forming large stands in wet meadows and streamsides in mts, 900-7000 feet elevation, mainly in the Cascades, mostly E of the crest, WA to central CA, but with outlying populations E to ID.

IDENTIFICATION TIPS: Three glaucous-leaved, rhizomatous sedges that are common E of the Cascades can easily be confused, and the perigynium veins that distinguish them may not develop until perigynia are fully mature, perhaps in August. *Carex angustata* has 1-3 weak veins on each side, *C. nebrascensis* has 5+ strong veins, and *C. aquatilis* lacks veins. *Carex nebrascensis* tends to live at lower elevations and in drier regions though in equally wet habitats. *Carex aquatilis* perigynium "shoulders" are rounder; those of *C. angustata* and *C. nebrascensis* are more sloping; only *C. nebrascensis* has distinct beak teeth.

COMMENTS: *Carex angustata* is frequently misidentified and overlooked in montane wet sedge meadows. Livestock and elk graze it. After fire, it readily regrows from deep rhizomes. In meadows of the Sierra Nevada, its habitats experience less annual fluctuation in water table than do those of *C. nebrascensis*. *Carex angustata* is apparently beginning to spread along roadsides into habitats where it did not historically occur. Due to its long, strong rhizomes, *C. angustata* has considerable potential for stabilizing disturbed streamsides, wet meadows and roadside ditches.

top left: perigynia
top right:
    inflorescences
center right:
    perigynium
bottom right: habit

Carex anthoxanthea J. Presl & C. Presl
Common name: Grassyslope Arctic Sedge
Section: *Circinatae*   Key: A

KEY FEATURES:
- Single spike
- Long, narrow perigynia
- Seepy, north-facing slopes on the Olympic Peninsula

DESCRIPTION: Habit: Rhizomatous with shoots arising singly, sometimes forming clumps. **Culms**: 5-40 cm long, delicate, longer than or as long as the leaves. **Leaves**: Flat, 1.5-2.5 mm wide, the lowest leaves reduced to bladeless sheaths. **Inflorescences**: One spike per culm, 1-2.7 cm long, lacking inflorescence bracts; each spike usually with only ♂ or ♀ flowers, not both, but rarely spikes androgynous. ♀ **scales**: Persistent, shorter than or as long as the perigynia, leaving the beak and upper part of the body exposed. **Perigynia**: Ascending, linear-lanceolate, with several veins as well as 2 ribs, 3-4.3 mm long, 0.8-1.1 mm wide, 3-4 times as long as wide, the distal margins smooth, beak tip dark. Stigmas (2-) 3. **Achenes**: Trigonous (or lenticular).

HABITAT AND DISTRIBUTION: One known PNW population in a cool, north-facing seep on the Olympic Peninsula of WA, at 2800 feet; elsewhere fens, bogs, muskegs, and wet meadows, 30-3000 feet, AK to WA, also Russian Far East.

IDENTIFICATION TIPS: In the PNW, *C. anthoxanthea* forms loose clumps in cool, moist habitat. Its narrow, solitary spikes contain distinctively narrow perigynia. The only species with similarly narrow mature perigynia are *C. circinata* and *C. pauciflora*. *Carex circinata* is cespitose with involute leaves and longer perigynia that have finely serrulate distal margins and hyaline beak tips. *Carex pauciflora* has deciduous ♀ scales and much longer, spreading to reflexed perigynia. Descriptions may imply that ♂ *C. anthoxanthea* and ♀ *C. scirpoidea* are similar, but their overall appearance is quite different, with *C. scirpoidea* more robust.

COMMENTS: The one known *C. anthoxanthea* population in WA is doubtless a remnant of larger populations present during the Ice Ages. It is legally protected but remains vulnerable because a road cuts through its cliff-side habitat. N of the PNW, this species forms large open mats or comes up as seemingly solitary shoots in a turf of mixed graminoids. In its PNW location, suitable substrate is patchy, and its rhizomes cross back and forth in the limited spaces, producing shoots in clusters and making the plant appear cespitose.

top left: perigynia   *top right*: inflorescences   *center*: habit   *bottom*: habitat

Carex aperta Boott in J.W. Hook.
Common name: Columbia Sedge
Section: *Phacocystis*   Key: G

KEY FEATURES:
- Perigynium inflated and olive, purplish, or orange in color
- Rhizomatous, forming large monocultures

DESCRIPTION: Habit: Rhizomatous but the deep rhizomes often not collected. **Culms**: 15-90 cm. Plant bases reddish brown. **Leaves**: Green. Leaf sheath fronts hyaline, usually with pale brown spots, not ladder-fibrillose. **Inflorescences:** Lowest inflorescence bract ± as long as the inflorescence. Lateral 2-3 spikes ♀ (or androgynous), erect, 1.5-3.5 cm long, 4-6 mm wide. Terminal 1-2 spikes ♂. ♀ **scales:** Reddish brown, acute, longer or shorter than the perigynia. **Perigynia**: Coppery brown, orangish, olive brown, or purplish with reddish brown spots on distal half, elliptic to obovate, veinless, inflated (i.e., larger than the achene and thus empty inside near the top), 2.5-2.8 mm long, 1.5-2 mm wide, only loosely enclosing the achene. Beak 0.1-0.3 mm, not bidentate. Stigmas 2. **Achenes**: Lenticular.

HABITAT AND DISTRIBUTION: Montane bogs, floodplains, lake shores, pond margins and wet sedge meadows, mainly in the Cascades but also in wet prairies in the Willamette Valley, OR, Puget Trough, WA, along the Columbia River, and scattered populations in NE OR and E WA. BC to MT and OR.

IDENTIFICATION TIPS: *Carex aperta* forms large, monospecific stands and is distinguished by its inflated perigynia in an odd shade of olive-green, purplish, or orangish. It lacks veins on the perigynia and therefore can be confused with *C. aquatilis* or *C. scopulorum,* which do not have inflated perigynia. Also, *C. scopulorum* grows at higher elevations and tends to have shorter spikes and has green or purple-black perigynia.

COMMENTS: *Carex aperta* was once a community dominant along the lower Columbia River, where it was harvested for hay. Populations have been greatly reduced by hydrologic changes associated with Columbia River dams, and by farming and development, though large stands remain at some sites, e.g., on Sauvie Island, near Portland. Populations in the Cascades appear secure. *Carex aperta* has considerable potential for restoration of disturbed wetlands in the Portland area and the Willamette Valley. This is one of the few plants that can persist in wetlands dominated by Reed Canarygrass (*Phalaris arundinacea*).

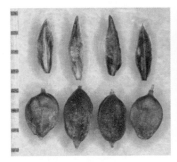

top left: pistillate scales, perigynia
top middle: spike
top right: inflorescence
bottom left: habitat
bottom right: habit

Carex aquatilis Wahlenb. var. *aquatilis*
Common name: Water Sedge
Section: *Phacocystis*   Key: G

KEY FEATURES:
• Perigynia veinless, green or tawny, often with brownish dots
• Plant bases chestnut, reddish brown, or yellowish brown
• Rhizomatous
• Lateral spikes erect

DESCRIPTION: Habit: Rhizomatous but the deep rhizomes often not collected. **Culms**: 20-120 cm. Plant bases reddish brown or brown. **Leaves**: Green or sometimes glaucous, 2.5-8 mm wide. Leaf sheath fronts not ladder-fibrillose. **Inflorescences:** Lowest inflorescence bract usually longer than the inflorescence. Lateral 2-6 spikes ♀, erect, not tapering at the base, 1-10 cm long, 3-7 mm wide. Terminal 1-4 spikes ♂. ♀ **scales**: Reddish brown or black, longer than the perigynia. **Perigynia**: Pale brown with reddish or brownish spots, elliptic to obovate, with rounded "shoulders," veinless or nearly so, 2-3.6 mm long, 1.3-2.3 mm wide. Beak up to 0.2 mm, not bidentate. **Achenes**: Lenticular, 1.1-1.8 mm long, 0.7-1.5 mm wide. Stigmas 2.

HABITAT AND DISTRIBUTION: Forming large stands in wet sedge meadows, bogs, streamsides, and lakeshores, in areas of slow-moving water where soil is inundated in early summer and has moisture in the root zone throughout the year; mainly E of the Cascade crest in WA and OR. Introduced to Coos Co., OR. AK to NF, S to NV, WI, and MI, also N Europe.

IDENTIFICATION TIPS: The chestnut, reddish brown, or yellowish brown plant bases distinguish *C. aquatilis* from most similar rhizomatous sedges. Similar *C. nebrascensis* has 5+ strong perigynium veins on each side when the perigynia are fully mature, and generally grows at lower elevations, in the Great Basin and the high lava plains. Both can have glaucous or green, V-shaped or W-shaped leaves. See *C. aquatilis* var. *dives*, which grows W of the Cascades and has long drooping lateral spikes, and also *C. angustata*.

COMMENTS: *Carex aquatilis* has long, strong rhizomes, can establish after disturbance and persist as vegetation matures, and is resistant to vehicular traffic, moderate grazing, fire, heavy metal contamination, and oil spills. It therefore has potential for wetland habitat restoration and for roadside plantings. Plugs may be more successful than seed at establishing populations. Seeding with non-native grasses interferes with establishment and spread of *C. aquatilis*.

top left: pistillate
  scales,
  perigynia,
  achenes
top right:
  inflorescence
center left: habit
bottom: habitat

Carex aquatilis Wahlenb. var. *dives* (T. Holm) Kük.

Common name: Sitka Sedge, Water Sedge
Section: *Phacocystis*   Key: G
Synonyms: *C. sitchensis* Prescott ex Bongard

KEY FEATURES:
- Perigynia veinless, green or tawny, often with purplish brown dots
- Plant bases chestnut, reddish brown, or yellowish brown
- Rhizomatous
- Long, drooping lateral spikes

DESCRIPTION: Habit: Rhizomatous but the deep rhizomes often not collected. **Culms**: 35-150 cm. Plant bases reddish brown, chestnut, or brown. **Leaves**: Green or glaucous, 5-18 mm wide. Leaf sheath fronts red-dotted but not densely so, not coppery, not ladder fibrillose. **Inflorescences**: Lowest inflorescence bract usually longer than the inflorescence. Lateral 2-6 spikes ♀ or androgynous, drooping, often tapering at the base, 4.5-11.5 cm long, 4-18 mm wide. Terminal 1-4 spikes ♂. **♀ scales**: Reddish brown or black, longer than the perigynia. **Perigynia**: Green or tawny with purplish brown spots, elliptic to obovate, veinless or nearly so, 1.9-3.5 mm long, 1-1.2 mm wide. Beak 0.3-0.4 mm, not bidentate. Stigmas 2. **Achenes**: Lenticular, 1.1-1.8 mm long, 0.7-1.5 mm wide.

HABITAT AND DISTRIBUTION: Wet meadows, bogs, lakeshores, and streamsides, sometimes forming large stands, mostly W of the Cascade crest. AK to northern CA, reported from MT, disjunct in mts of NC.

IDENTIFICATION TIPS: Plant bases are chestnut, reddish brown, or yellowish brown, not purplish black as in other rhizomatous *Phacocystis*. *Carex aquatilis* var. *dives* may occur with *C. obnupta* and *C. lyngbyei,* both of which have very thick-walled, hard perigynia. *Carex aquatilis* var. *aquatilis* grows mainly E of the Cascades and has shorter, erect spikes. Many *C. aquatilis* populations, especially in the Cascades, have plants with intermediate morphology or even both extremes, and cannot be identified as one variety or the other.

COMMENTS: Long, deep rhizomes of *C. aquatilis* var. *dives* can extend under channels of small headwater streams, tying the banks together and preventing downcutting. Deep rhizomes survive moderate fires. Foliage is readily grazed by livestock and elk; plants are resistant to moderate grazing, but are weakened by prolonged heavy grazing. Humans can eat the succulent stem bases raw. Native Americans of the PNW used the tough, fibrous leaves to make strong handles for baskets and woven bags.

top left: pistillate scales, perigynia
top right: typical inflorescence
center left: inflorescence tending toward
   typical *C. aquatilis* var. *aquatilis*
bottom: habitat

Carex arcta Boott
Common name: Northern Clustered Sedge
Section: *Glareosae*   Key: I

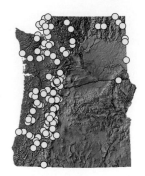

KEY FEATURES:
- 5-15 crowded, gynecandrous spikes
- Perigynia not winged, green to brown, beaked
- Meadows and thickets

DESCRIPTION: Habit: Cespitose. **Culms**: 15-80 cm long. **Leaves**: 2-4 mm wide, green, shorter to longer than the culms; leaf sheath fronts with tiny purplish dots. **Inflorescences**: Usually green to light brown, 1.5-4 cm long, 7-12 mm wide, with 5-15 crowded spikes. Spikes gynecandrous, oblong, 5-10 mm long, with 10-20 perigynia, the beak tips giving the spikes a somewhat rough outline. ♀ **scales**: Hyaline and often brownish, with green midvein, shorter than the perigynia. **Perigynia**: Spreading-ascending, green to brown, with several veins, ovate, 2-3 (-3.5) mm long, 1.2-1.5 mm wide. Beak 0.75-1.25 mm long. Dorsal suture conspicuous. Stigmas 2. **Achenes**: Lenticular.

HABITAT AND DISTRIBUTION: Wet places in meadows, woodlands, thickets, and forest openings, from near sea level to near timberline; mainly in and W of Cascades, also mts of N WA and NE OR. AK to ME, south to CA and MA, but not in the Great Plains.

IDENTIFICATION TIPS: *Carex arcta* is usually identified by elimination. Perigynia are not winged, so it can't be a member of section *Ovales*. *Carex canescens* has gray-green foliage and fewer (4-8) spikes per inflorescence, and looks much "neater"; its spikes have a smoother outline due to its shorter perigynium beaks.

COMMENTS: *Carex arcta* can be a community dominant in wetlands in the southern Cascades of OR, but is more often seen in small populations. It is rare in most of the eastern part of its range, and uncommon to rare in northern WA.

Carex arcta has been reported to have 54, 58, or 60 chromosomes per cell. *Carex* tend to have many small chromosomes and variable chromosome numbers, such as 20, 22, 24, 30, and 36 in *C. deflexa* var. *deflexa*, probably because of their holocentric chromosomes which attach to spindle fibers along their length, not just at one point (the centromere). This trait is shared by a few species of protista, plants, and animals such as nematode worms, bees, wasps, and some ticks. If holocentric chromosomes break, the pieces need not be lost during cell division. Rather, each piece can function as a chromosome. In contrast, broken human chromosomes result in loss of part of the DNA, often causing birth defects.

top left: perigynia
top right:
   inflorescence
bottom left: habit
bottom right:
   inflorescences

Carex arenaria L.
Common name: Sand Sedge
Section: *Ammoglochin*   Key: main

KEY FEATURES:
- Rhizomatous patches
- Perigynia winged but spikes ♀ or androgynous
- Sandy soils near Portland

DESCRIPTION: Habit: Rhizomatous, forming large populations, the rhizomes with a thick pithy cortex. **Culms**: 15-60 cm tall, erect. **Leaves**: 1.5-4 mm wide; top of leaf sheath front yellowish, thickened, concave. **Inflorescences**: 2-8 cm long, consisting of 5-15 overlapping, ascending spikes, somewhat shaggy-looking. Spikes androgynous or ♀ (or the uppermost ♂). ♀ **scales**: Reddish brown with green midrib and hyaline margins, usually longer than the perigynia. **Perigynia**: Light brown, lanceolate to broadly ovate, planoconvex, several veined, (4-) 4.5-6 mm long, 1.3-3.3 mm wide, broadly winged above the middle. Beak 1.2-2.5 mm long. Stigmas 2. **Achenes**: Lenticular.

HABITAT AND DISTRIBUTION: Introduced to sandy shores of rivers in Portland, OR. In Europe, where it is native, habitat is mainly moving sands near the coast but also inland sandy soils, usually in full sun but persisting in shade, e.g., in pine forests. Introduced to E coast of N America, DE to NC, and to Australia.

IDENTIFICATION TIPS: *Carex arenaria* is a strongly rhizomatous sedge of sandy soils, with shaggy inflorescences and large, partly winged perigynia. Its winged perigynia suggest *Carex* section *Ovales,* which are cespitose with gynecandrous spikes. The shaggy inflorescences of *C. arenaria* somewhat resemble *C. douglasii,* which has smaller heads and smaller perigynia and grows E of the Cascades.

COMMENTS: *Carex arenaria* was introduced to Portland, probably in ship ballast sand. Collected in 1916, it was rediscovered on Sauvie Island in 2003. *Carex arenaria* is a model organism for study of ecology and physiology of clonal plants. A rhizome segment can live for 10 years, its shoots dying after the second year but the rhizome remaining active in transport of water and nutrients to younger shoots. "Sinker roots" at each shoot base may extend 10 feet to water. Fine, dense, shallow roots explore soil near the rhizome for nutrients and water. Roots are colonized by mycorrhizal fungi. Plants are tolerant of drought but vulnerable to competition from other plants, and to trampling. Viable seed has been found in feces of pheasants. In Europe, humans have eaten the seeds and starchy rhizomes. A decotion from the rhizomes has been used to treat diverse ailments.

Carex arenaria

top left: perigynia
top right: inflorescences
center left: habitat
bottom: habit

Carex atherodes Spreng.
Common name: Awned Sedge
Section: *Carex*   Key: C, E

KEY FEATURES:
- Perigynia with long, spreading teeth
- Leaves and top of leaf sheath front hairy
- Lake margins, marshes, and ditches E of Cascades

DESCRIPTION: Habit: Rhizomatous, forming large stands. **Culms**: 35 –200 cm tall. **Leaves**: Usually sparsely hairy, 3-10 mm wide. Basal leaf sheaths reddish purple. Leaf sheath front usually pubescent at least near the top, becoming ladder-fibrillose with age. Leaves and sheaths that develop underwater may be glabrous. **Inflorescences**: Lateral 3-6 spikes ♀, 2-10 cm, erect or ascending; terminal 2-6 spikes ♂, 2-6 cm. ♂ **scales**: With scabrous awn, glabrous; rarely sparsely pubescent. ♀ **scales**: Narrow, glabrous or scabrous along midvein, with scabrous awn. **Perigynia**: With sparse spreading hairs (occasionally glabrous), and with 12-20 veins, (6.5-) 7-12 mm long, 1.8-3.8 mm wide. Beak 2.1-4 mm long, with spreading teeth (1.2-) 1.5-3 mm long. Stigmas 3, style persistent, straight. **Achenes**: Trigonous.

HABITAT AND DISTRIBUTION: Lake margins, meadows, marshes, streams, and ditches, often alkaline, generally in sites with water as deep as 2.5 feet in spring and remaining wet to moist in summer; E of the Cascades in WA and OR. AK to ME, south to CA, NM, and VA, also Eurasia.

IDENTIFICATION TIPS: *Carex atherodes* is a tall sedge forming stands in shallow water, with usually hairy foliage and perigynia, and long, spreading beak teeth. The similar *C. sheldonii* may grow on adjacent drier terraces but is less robust with narrower leaves, shorter, more densely pubescent perigynia, and shorter beak teeth. *Carex utriculata* and *C. lacustris* have glabrous foliage, straight beak teeth, and contorted styles.

COMMENTS: *Carex atherodes* is uncommon in the PNW, but where present may form extensive stands. *Carex atherodes* is an important food for American Bison and a good source of protein for Muskrats. It is cut for hay in OR.

top left: pistillate
   scale
top middle:
   perigynium
top right:
   inflorescence
center left: hairy leaf
   sheath front
center middle:
   ladder-fibrillose
   leaf sheath
   front
center right: spikes
bottom: habitat

Carex athrostachya Olney
Common name: Long-bract Sedge
Section: *Ovales*   Key: J

KEY FEATURES:
- Cespitose, with gynecandrous spikes and winged perigynia
- Lowest inflorescence bract usually longer than the inflorescence, bristle-like
- Inflorescence straw-colored or light tan
- Disturbed and/or vernally wet areas

DESCRIPTION: Habit: Cespitose. **Culms**: (5-) 20-80 cm tall. **Leaves**: 10-20 cm long, (1.5-) 2-3 (-5) mm wide. **Inflorescences**: (0.8-) 1.5-2.2 cm long, 7-20 mm wide, dense and head-like, with lowest 1-3 inflorescence bracts usually longer than the inflorescence, bristle-like, sometimes leaf-like. Spikes gynecandrous. ♀ **scales**: Gold to brown with midrib paler, green, or the same color, 2.4-4.3 mm long, shorter and narrower than the perigynia. **Perigynia**: Cream-colored to light brown, lanceolate, (2.8) 3.5-4 (-4.8) mm long, (0.8) 1-1.5 (-1.8) mm wide, with 0-9 veins on each face, with wing (0.1-) 0.2 (-0.5) mm wide. Beak tip gold to reddish brown, unwinged, and parallel-sided for the distal 0.4-0.9+ mm; 1.9-2.5 mm from top of achene to tip of beak. Stigmas 2. **Achenes**: Lenticular, (1-) 1.2-1.6 mm long, 0.7-1 mm wide, 0.3-0.4 mm thick.

HABITAT AND DISTRIBUTION: Seasonally wet places, disturbed sites, pond margins, reservoir margins, roadside puddles, and riparian areas, mainly in and E of the Cascades, occasionally on the W side, perhaps as a recent introduction. Low to high elevations. AK to MB, S to CA and TX.

IDENTIFICATION TIPS: *Carex athrostachya* can usually be identified by its lowest inflorescence bract, which is usually longer than the inflorescence but not leaf-like. *Carex unilateralis* has long, usually leaf-like bracts, the head angled as if growing from side of stem, and perigynia that are usually longer and flat and ciliate-serrulate to the tip. Occasionally *C. athrostachya* plants lacking long inflorescence bracts are identified by habitat and by their fine-textured, light-colored, dense heads with small, narrow perigynia. Late in the season, many sedge species put out a second set of inflorescences that have unusually long subtending bracts that make them resemble *C. athrostachya*. See discussion of rare *C. crawfordii*.

COMMENTS: *Carex athrostachya* is a common sedge of disturbed moist or seasonally moist sites. In heavily disturbed sites, it may flower in the first year and thus function as an annual. It increases in moderately disturbed sites and after fire. In cultivation, it grows well and produces abundant seed, and is useful for habitat restoration projects.

top left: perigynia
top right: infllorescence
center left and right: infllorescences
bottom: habit

Carex atrosquama Mack.

Common name: Brass-fruit Sedge
Section: *Racemosae*   Key: F
Synonyms: *C. atrata* L. ssp. *atrosquama* (Mackenzie)
   Hultén, *C. atrata* L. var. *atrosquama* (Mackenzie)
   Cronquist

KEY FEATURES:
- Perigynia dark gold with darker beak tip
- ♀ scales obtuse, dark purplish black
- Subalpine to alpine meadows and shorelines

DESCRIPTION: Habit: Cespitose. **Culms**: 20-50 cm tall. **Leaves**: 3-5 mm wide. **Inflorescences:** 3-4 (-6) spikes, each 0.8-2 cm long, the lateral spikes ♀ and erect, terminal spike gynecandrous. ♀ **scales**: Dark purplish black, usually shorter than the perigynium, tip obtuse, the midrib usually dark like the rest of the scale or at least not strongly contrasting with it. **Perigynia**: Elliptic, more or less flattened except over the achene, green when young but maturing dark gold with darker tip, 2.5-3.5 mm long, 1.5-1.75 mm wide, papillose distally, with a small beak 0.3-0.5 mm long. Stigmas 3. **Achenes**: Trigonous, nearly filling bodies of perigynia.

HABITAT AND DISTRIBUTION: Alpine and subalpine grasslands, moist to mesic meadows, and streamsides in N WA and NE OR. AK to NT, S to OR, UT, and CO.

IDENTIFICATION TIPS: *Carex atrosquama* is a cespitose, high elevation plant with dark gold, papillose, elliptical perigynia mostly hidden by blackish ♀ scales. The perigynium papillae can be seen at 10X. *Carex heteroneura* can be similar but has brown, elliptic to obovate perigynia with a smooth surface. (Perigynium cell walls form a network of low ridges visible at high magnification, but lack papillae). Also, its spikes are usually longer.

COMMENTS: *Carex atrosquama* is rare in the PNW and is uncommon or local in much of its range. Like *C. heteroneura,* this species has been treated as a component of *C. atrata. Carex atrosquama* is consistently different from *C. heteroneura*.

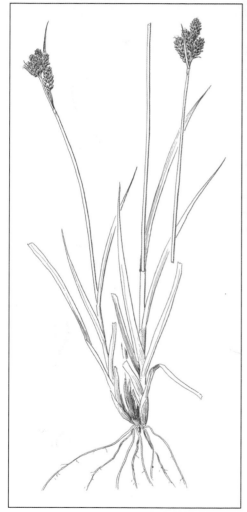

top left: perigynia
top right: inflorescences
left: habit

Carex aurea Nutt.

Common name: Golden Sedge
Section: *Bicolores*   Key: G

KEY FEATURES:
- Mature perigynia globose, succulent, orange
- ♀ scales deciduous
- Plants short, with delicate, arching leaves

DESCRIPTION: Habit: Rhizomatous, the rhizomes about 1 mm in diameter. **Culms**: 5-40 cm tall. **Leaves**: Pale green to slightly glaucous, 3-20 (-40) cm long, (1.4) 2-3.5 mm wide. Plant bases whitish or medium brown. **Inflorescences**: Lateral spikes ♀, 0.4-2.0 cm long, 3-5 mm wide, with perigynia relatively loosely spaced within the spike, with middle internodes (0.2-) 0.5-1.5 mm, averaging 0.65 mm. Terminal spike usually ♂ (rarely gynecandrous), when ♂ usually 0.5-1.7 cm long, averaging 0.9 cm long, 2 mm wide. ♀ **scales**: 1.2-2.5 mm, obtuse to acute, sometimes awned, often falling before the perigynia. **Perigynia**: Globose or obovate, nearly veinless at maturity but veined when young, 1.6-3.2 mm long, 1.1-2 mm wide. Fresh mature perigynia pale greenish becoming orange and more or less succulent when ripe, sometimes white-powdery. Dried mature perigynia dark brown, slightly waxy, and squashed, like dried chokecherries. Immature perigynia green to tan, often white-powdery. Perigynium beakless or nearly so, the perigynium apex sometimes curved to one side. Stigmas 2. **Achenes**: Lenticular, 1.3-1.8 (-2) mm long, 1-1.6 mm wide.

HABITAT AND DISTRIBUTION: Common but inconspicuous in springs, riparian areas, or wet meadows, often an understory plant under taller grasses or sedges; throughout WA and OR. Low to high elevations, but restricted to middle or high elevations in the southern part of its range. Throughout N America except the SE.

IDENTIFICATION TIPS: *Carex aurea* is a small sedge with succulent yellow, orange, or sometimes whitish mature perigynia loosely spaced in the spike. ♀ scales fall early, often before the perigynia, and ♂ spikes are small. Distinctive ripe perigynia fall from plant, making it difficult to collect good specimens. *Carex hassei* is similar and not distinguishable when young. No single trait works to distinguish *C. aurea* from all *C. hassei*, so until recently PNW botanists treated them all as part of a variable *C. aurea*. *C. hassei* ♂ spikes average longer and wider, its perigynia are never succulent throughout (but may be so at base), they are sometimes crowded, and its ♀ scales are sometimes persistent.

COMMENTS: The orange, succulent, mature perigynia may be eaten and dispersed by birds, and are edible for humans.

top left: perigynia
top right : inflorescences
center left: inflorescences
bottom right: habit

Carex barbarae Dewey
Common name: Whiteroot Sedge
Section: *Phacocystis*   Key: G

KEY FEATURES:
- Rhizomatous, with purplish black plant bases
- ♀ scales with scabrous awns
- Along Rogue River and its major tributaries

DESCRIPTION: Habit: Rhizomatous, but deep rhizomes often not collected; forming large stands. **Culms**: 30-110 cm. **Leaves**: Green to somewhat glaucous, 4-9 mm wide. Plant bases reddish brown or purplish black. Leaf sheath fronts ladder-fibrillose, coppery colored, densely red-dotted. **Inflorescences**: Lowest inflorescence bract longer than the inflorescence. Lateral 3-6 spikes ♀, brown, erect to drooping, tapering at base, 3-10 cm long, 5-6 mm wide. Upper 2-3 spikes ♂. **♀ scales**: Reddish brown, acute, with a scabrous awn up to 2 mm long. **Perigynia**: Brown with reddish spots, somewhat tough but not hard, faintly veined, 3-4 mm long, (1.9) 2.2-2.5 mm wide. Beak ~ 0.5 mm, bidentate, scabrous. Stigmas 2. **Achenes**: Lenticular.

HABITAT AND DISTRIBUTION: Forming large stands in floodplains and terraces, in open woodlands, savannas, meadows, and ditches along the Rogue River and its major tributaries; SW OR to S CA.

IDENTIFICATION TIPS: *Carex barbarae* is a highly variable species most likely to be confused with *C. obnupta* with which it sometimes grows in mixed populations. *Carex obnupta* has hard perigynia and blackish inflorescences, and lacks awns on the ♀ scales. *Carex aquatilis* var. *dives* usually has longer spikes and usually lacks awns on the ♀ scales. Some odd, possibly hybrid populations occur in the Rogue River watershed.

COMMENTS: Historically, more than a third of California's Native American tribes used the rhizomes and split leaves of *C. barbarae* for weaving baskets. They thinned and weeded *C. barbarae* beds in open riparian woodland to encourage growth of long, straight rhizomes. Coils of dried, peeled, split rhizomes were important articles of trade. Habitat destruction, succession, and invasive plants have greatly restricted *C. barbarae*, but a few continuously tended beds exist in California. *Carex barbarae* can stabilize soil well and has been recommended for planting on roadsides in appropriate habitat, but it can be aggressive. *Carex barbarae* seems unusually variable over its range, perhaps a result of humans transporting preferred plants over long distances. *Carex barbarae* may have originated from hybridization, perhaps between *C. obnupta* and *C. nebrascensis*.

top left: perigynium, pistillate scale
top middle: perigynia
top right: inflorescence
center left: inflorescences
bottom: habit, habitat

Carex bebbii (L. H. Bailey) Olney ex Fernald
Common name: Bebb's Sedge
Section: *Ovales*   Key: J

KEY FEATURES:
- Cespitose, with gynecandrous spikes and winged perigynia
- Inflorescence of crowded, roundish, red-brown spikes
- Perigynia flat and ciliate-serrulate almost to the tip
- Wet meadows, ditch margins

DESCRIPTION: Habit: Cespitose. **Culms**: 20-90 cm tall. **Leaves**: 11-25 cm long, 1.7-4.2 mm wide. **Inflorescences**: 1.1-3 cm long, 5-14 mm wide, with 3-10 overlapping, more or less round spikes that can be distinguished. Spikes gynecandrous. ♀ **scales**: Reddish brown, with midstripe that may be green, pale, or brown, 2.5-3.5 mm long, shorter and narrower than their perigynia. **Perigynia**: Spreading, green when young, becoming reddish brown, ovate or elliptic, planoconvex, 2.5-3.8 mm long, (1-) 1.2-2 mm wide, with 3+ dorsal veins, the ventral side with 1-3 faint veins or veined only near the base, with wings 0.2-0.5 mm wide. Beak tips reddish brown, winged, and more or less ciliate-serrulate; 1.2-2.2 mm from top of achene to tip of beak. Stigmas 2. **Achenes**: Lenticular, 1-1.3 mm long, 0.6-0.9 mm wide, 0.3-0.4 mm thick.

HABITAT AND DISTRIBUTION: Wet meadows, stream banks, irrigation ponds, roadside ditches, generally in areas with basic to neutral pH; E of the Cascade crest, also mts of N WA. AK to NF, S to OR, CO, and MA.

IDENTIFICATION TIPS: *Carex bebbii* is one of those species that is difficult to key but relatively easy to recognize once you know it. The reddish brown mature inflorescences with their roundish spikes are good clues. It is most similar to *C. subfusca,* which also has small perigynia but its inflorescences are green, straw-colored, or drab light brown and its perigynia often have beak tips unwinged and entire for 0.4-0.7 mm. *Carex pachystachya* may grow in the same habitat as *C. bebbii* but its inflorescences are darker and more crowded, and its perigynia beaks are unwinged, brown, and entire for 0.4-0.7 mm from the tip.

COMMENTS: *Carex bebbii* may mature and flower in the first growing season, thus acting like an annual. In OR it has been considered rare, but it appears to be exploiting human disturbance to expand its range along roadsides and irrigation ditches. This species has been used in habitat restoration projects, rainwater control ponds, and gardens. Most commercially available seeds and plugs originate from the NE U.S. and should not be used in the PNW.

Carex bebbii

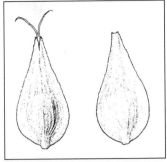

top left and immediately below:
  perigynia
top right: inflorescence
center left: inflorescences
bottom: habit

Carex bolanderi Olney

Common name: Bolander's Sedge
Section: *Deweyanae*   Key: I
Synonyms: *C. deweyana* Schwein. ssp. *leptopoda*
  (Mack.) Calder & Taylor, *C. deweyana* Schwein var.
  bolanderi (Olney) W. Boott

KEY FEATURES:
• Loosely cespitose, with lax, grass-like leaves
• In moist forest
• Thin, membranous perigynia with long beaks and
  long teeth
• Ligules longer than wide

DESCRIPTION: Habit: Cespitose. **Culms**: 31-115 cm long, smooth or minutely antrorsely scabrous (at 20X) at mid-length. **Leaves**: 1.8-5.9 mm wide, with ligule of uppermost culm leaf much longer than wide, (2-) 3.5-7.1 mm long. **Inflorescences**: Spikes 5-9, light gold to straw-colored (to green), relatively rough in outline; usually gynecandrous (or ♀), lowest spikes usually longer than wide, 12-22 (-25) mm long, with 14-30+ perigynia; terminal spike 9.6-16 mm long. ♀ **scales**: gold or straw-colored (or white) with green midvein, usually awned (awn to 1.5 mm), the body about as long as or longer than the mature perigynium body. **Perigynia**: Green, straw-colored, to reddish brown, 3.4-5.2 mm long, with beak 1.5-2.5 mm long, (38-) 40-50+% of perigynium length, with teeth (0.2-) 0.3-1 mm long. Stigmas 2. **Achenes**: Lenticular. **Anthers**: 1.3-1.8 mm long.

HABITAT AND DISTRIBUTION: Shaded to partially shaded riparian zones and moist forest, especially E of Cascades and in SW OR but scattered throughout the PNW. BC to MT, S to CA and NM.

IDENTIFICATION TIPS: Several characteristics can be used to distinguish the "identical triplets" *C. bolanderi, C. infirminervia,* and *C. leptopoda. Carex bolanderi* is the one with long, often spreading beak teeth. Do not be misled by perigynia that were torn between the teeth, making them seem longer. *Carex infirminervia* has shorter beak teeth. *Carex leptopoda* has white to greenish ♀ scales, shorter perigynium beaks, shorter teeth, and spikes with a smoother outline. In all three species, the beak matures before the perigynium body, making beak/perigynium proportions misleading on young material.

COMMENTS: *Carex bolanderi* is the member of the *C. deweyana* complex most likely to be found in isolated riparian zones in shrub steppe or dry, open pine forest. *Carex bolanderi* seems to mature later in the summer than *C. leptopoda,* which may keep the two genetically distinct where they grow together. See notes under *C. infirminervia* and *C. leptopoda.*

top left: perigynia
top right: inflorescences
center left: ligule
bottom: habit

Carex brainerdii Mack.
Common name: Brainerd's Sedge
Section: *Acrocystis*   Key: C

KEY FEATURES:
* Perigynia pubescent and veined
* Basal ♀ spikes
* Cespitose; foliage usually glaucous

DESCRIPTION: Habit: Densely to loosely cespitose. **Culms**: 5-35 cm tall. Plant bases reddish. **Leaves**: Glaucous or seldom green, usually longer than the culms, 1.8-3.8 mm wide, densely papillose on the lower surface. **Inflorescences**: Bracts of lowest lateral spikes leaf-like, as long as or longer than the inflorescence. Lateral spikes (just below the terminal spike) 2-6, ♀, 4-8 mm long, with 1-4 perigynia. Terminal spike ♂, 0.6-1.3 cm long. There are also 1-4 basal ♀ spikes ± hidden among the leaf sheaths at the base of the plant. **♀ scales**: Reddish brown with or without narrow white margins, shorter than the perigynia, apex acute to acuminate. **Perigynia**: Pubescent, elliptic to obovoid, with succulent bases that wither when dry, with 12-15 veins to at least mid-body as well as 2 ribs, 4-5.3 mm long, 1.4-2.2 mm wide; beaks 0.8-1.9 mm long, with apical teeth 0.2-0.5 mm long. Stigmas 3. **Achenes**: Trigonous.

HABITAT AND DISTRIBUTION: Uplands, including open conifer forests, dry openings, rocky areas, roadsides, cut banks, and clearcuts, generally on excessively drained soils, sometimes on serpentine, often in disturbed areas, at moderate to high elevations; SE Lane Co., OR, S to CA.

IDENTIFICATION TIPS: *Carex brainerdii* forms loose, usually glaucous clumps in dry, more or less open habitats. Its lower leaf surfaces are densely papillose at 20X. Like *C. rossii* and *C. deflexa* var. *boottii,* it has pubescent perigynia, some of them in basal spikes nestled among the leaf sheaths, and it blooms very early in May or June. *Carex rossii* and *C. deflexa* have perigynia that are veinless on the faces and narrower, usually green leaves with only occasional papillae. Leaves of some SW OR *C. rossii* are glaucous but lack dense papillae. *Carex inops* ssp. *inops* lacks basal ♀ spikes and has a more open, long-rhizomatous growth form.

COMMENTS: *Carex brainerdii* can grow in erosion-prone areas and hold soil well because its root system is more extensive than its size might suggest. However, use in erosion control has been limited by its low seed yield and its short stature, which makes harvest difficult. Existing plants can be divided to produce plugs for planting. *Carex brainerdii* has value as a garden subject in well-drained soils.

top left: perigynia
top right: inflorescence
bottom: habit

Carex brevicaulis Mack.
Common name: Short-stemmed Sedge
Section: *Acrocystis*   Key: C
Synonyms: *C. deflexa* Hornem. var. *brevicaulis*
  (Mackenzie) Boivin, *C. rossii* Boott, misapplied

KEY FEATURES:
* Pubescent perigynia
* Basal ♀ spikes
* Coastal, sandy soils

DESCRIPTION: Habit: Cespitose or becoming rhizomatous with age, the rhizomes branching frequently to form spreading turf. **Culms**: 2-20 cm tall. Plant bases reddish. **Leaves**: Green to yellowish green, usually longer than the culms, 1.3-2 (-3.5) mm wide, arching and somewhat leathery, not densely papillose on the lower surface. **Inflorescences**: Bracts of lowest lateral spikes usually scale-like or bristle-like and shorter than the inflorescence, occasionally leaf-like and longer than the inflorescence; lateral spikes 2-5, ♀, 0.6-0.8 cm long, with 1-6 perigynia. Terminal spike ♂, 0.5-1.4 cm long. There are also 1-2 basal ♀ spikes ± hidden among the leaf sheaths. **♀ scales**: Pale to dark reddish brown with narrow white margins, apex acute to acuminate. **Perigynia**: Pubescent, globose, with succulent bases that wither when dry, with 2 ribs but otherwise veinless (or finely veined at the base); perigynia from lateral (not basal) spikes, 3.5-4.8 mm long, 1.7-2.1 mm wide with beaks 0.6-1.6 mm long, with apical teeth 0.2-0.5 mm long. Stigmas 3. **Achenes**: Trigonous.

HABITAT AND DISTRIBUTION: Stabilized sand dunes at the coast, generally in dry meadow-like plant communities, not with European Beach Grass; BC to CA.

IDENTIFICATION TIPS: This short sedge forms patches or turf of arching, leathery leaves in sandy soils. Like *C. rossii,* which may grow on coastal headlands, it produces pubescent perigynia in early spring and has basal pistillate spikes. *Carex rossii* has non-leathery leaves and a long, leafy inflorescence bract, and does not grow on sand dunes or form turf. *Carex pansa* lives on sand dunes but has glabrous perigynia in a single dense head, and grows in lines with rarely branching rhizomes.

COMMENTS: *Carex brevicaulis* habitat is threatened by succession to Shore Pine forest and by housing developments. Dune stabilization by European Beach Grass prevents formation of new habitat. Some populations thrive below the lawnmower in coastal picnic areas. Differentiating herbarium specimens from *C. rossii* and *C. deflexa* is so difficult that the taxa have often been synonymized. However, species status is supported by the distinctively tough, arching leaves that *C. brevicaulis* produces even in the greenhouse.

top left: perigynia
top right: inflorescences
center left and right: habit
bottom: habitat (stabilized coastal sand dunes)

Carex brevior (Dewey) Mack. ex Lunell
Common name: Plains Oval Sedge
Section: *Ovales*   Key: J

KEY FEATURES:
- Cespitose, with gynecandrous spikes and winged perigynia
- Perigynium body broadly winged, almost round
- Low-elevation meadows, ditches

DESCRIPTION: Habit: Cespitose. **Culms**: 15-120 cm tall. **Leaves**: 12-30 cm long, 1.5-3.5 mm wide. **Inflorescences**: (1.3-) 2.5-5 (-6.5) cm long, 5-10 (20) mm wide, the spikes overlapping but clearly distinguishable. Spikes gynecandrous. ♀ **scales**: White-hyaline with some brown color, midstripe often gold or green, lanceolate to ovate, shorter than the perigynium beaks, narrower than the perigynia, the tips usually acute. **Perigynia**: Green or straw-colored, the body orbicular or broadly ovate, planoconvex, (3-) 3.4-4.8 (-5.2) mm long, (2-) 2.3-3.2 mm wide, widest near the middle; lacking veins or sometimes with 1-5 faint dorsal veins, the wing 0.3-0.8 mm wide. Beak winged and ciliate-serrulate almost to the tip; 1.5-2.4 mm from top of achene to tip of beak. Stigmas 2. **Achenes**: Lenticular, 1.5-2.2 mm long, 1.3-1.8 mm wide, 0.5-0.6 mm thick.

HABITAT AND DISTRIBUTION: Mesic grasslands and prairies, sandy stream banks, open woods, road ditches, at low to moderate elevations; native in damp, grassy spots on banks of major rivers in E WA, introduced to a road ditch in Jackson Co., OR. BC to ME, south to OR, AZ, and GA, introduced to Japan.

IDENTIFICATION TIPS: With its broad wings that make the perigynium bodies almost round, *C. brevior* is relatively distinct. *Carex straminiformis* and *C. proposita* are broad-winged but grow on dry sites in high mountains. *Carex longii*, introduced to the coast, has perigynia broadest above the middle. *Carex leporina* has more ovate perigynium bodies to 2.1 mm wide. *Carex scoparia* has wide perigynium bodies, but with the wing narrowed below the middle, appressed perigynia, and fine-textured spikes with acuminate ♀ scales. *Carex molesta,* not yet in the PNW but introduced to CA, is extremely similar but its perigynia have 3-6 distinct veins over the achene on the ventral surface, while *C. brevior* has none or only faint ones that usually do not extend across the length of the achene.

COMMENTS: In eastern N America, *C. brevior* is an important component of prairie and mesic to dry grassland. In the PNW, *C. brevior* appears native only to a restricted habitat along major rivers in eastern WA. Revegetation projects in the PNW should use only seed collected within this limited PNW habitat.

top left: perigynia
top right: inflorescence
bottom left: habit
bottom right: inflorescences

Carex breweri Boott

Common name: Brewer's Sedge
Section: *Inflatae*   Key: A
Synonyms: *C. breweri* Boott var. *breweri*

KEY FEATURES:
- Single, fat spike
- Roundish, flat perigynia
- High elevation, well-drained soils in the Cascades

DESCRIPTION: Habit: Rhizomatous, the rhizomes 2-3 mm thick. **Culms**: 15-22 (-30) cm tall. **Leaves**: ± cylindrical, 0.6-1.2 mm wide, the leaf sheath fronts uniformly colored or often mottled with dark brown. **Inflorescences**: One thick, androgynous spike per culm, 1.4-2.5 cm long, 7-12 mm wide, lacking inflorescence bract. ♀ **scales**: Shorter and narrower than the perigynia, 3-5 veined, white down the center, the lower ones sometimes awned. **Perigynia**: Ascending, broadly elliptic, with 3-10 short veins on the faces, 5-7 mm long, 3-4.8 mm wide, with rachilla longer than the achene and the achene much smaller than the perigynium. Stigmas 3. **Achenes**: Trigonous.

HABITAT AND DISTRIBUTION: Dry, open slopes above 6000 feet, often on pumice; Mt. Adams, WA, S in the Cascades and Sierra Nevada to CA and NV.

IDENTIFICATION TIPS: Its almost round heads, short stature, and high elevation habitat make *C. breweri* distinctive. The only truly similar species is *C. engelmannii,* which differs in having ♀ scales that are about as long as the perigynia, with only one vein, and with a yellowish brown area around the midrib. It has shorter inflorescences and shorter, elliptic perigynia. Its rhizomes are narrower and its leaf anatomy differs. *Carex breweri* and *C. engelmannii* are mostly allopatric, but their ranges overlap on Mt. Adams, WA.

COMMENTS: An easy place to view *C. breweri* is in Crater Lake National Park on roadsides near Llao Rock. The broad, flat perigynia are wind-dispersed, like those of *C. straminiformis,* which occupies similar habitats nearby. Its odd, dense heads might make *C. breweri* an interesting garden subject.

top left: pistillate scales,
    perigynia
top right: inflorescences
bottom left: habitat
bottom right: habit

Carex brunnescens (Pers.) Poir. ssp. brunnescens
Common name: Brown Sedge
Section: *Glareosae*   Key: I

KEY FEATURES:
- Cespitose, with many slender, arching culms
- Inflorescence with very small, widely separated spikes
- Perigynia with short beaks with obvious dorsal sutures

DESCRIPTION: Habit: Cespitose. **Culms**: 15-60 cm long, erect or arching. **Leaves**: (1-) 1.5-2.5 mm wide. **Inflorescences**: Spikes 5-10, gynecandrous, with 5-10 ♀ flowers, 3-7 mm long, usually widely separated. ♀ **scales**: White-hyaline (tinged brown when growing in sun) with a green mid-stripe, about as long as the perigynia but not concealing them. **Perigynia**: Green to brown and usually dark brown at maturity, 2-2.5 mm long, with short beak 0.3-0.5 mm long. Dorsal suture extends onto the perigynium body, is about the same color as or lighter than the surrounding tissue, and may have a hyaline flap. Stigmas 2. **Achenes**: Lenticular.

HABITAT AND DISTRIBUTION: Seasonally wet spots in moist woodlands and thickets; in mountains of northern WA, also S in Cascades to Pierce Co. and Mt. St. Helens, WA; perhaps also Mt. Hood, OR. AK to Greenland, south to CA and GA, but absent in Great Plains.

IDENTIFICATION TIPS: *Carex brunnescens* is a densely cespitose plant with many leaves and arching culms, each with several widely separated, short spikes. Its perigynia have very short beaks and are dark brown at maturity. It is easily confused with young or small-fruited individuals of *C. laeviculmis,* which have proportionately longer perigynium beaks 0.4-1+ mm long, and ♀ scales that are straw-colored to reddish brown and shorter than the perigynia. Typical *C. canescens* differs in having gray-green foliage and perigynia. All *C. canescens* have dorsal sutures shorter than the beak. *Carex canescens* usually grows in more open habitats. Perigynia of *C. praeceptorum* resemble those of *C. brunnescens,* but the plants do not. *Carex praeceptorum* has shorter, erect culms with overlapping spikes.

COMMENTS: *Carex brunnescens* has been reported repeatedly from OR and parts of WA where it probably does not grow; the specimens have turned out to be *C. laeviculmis*. Continent-wide, this species is variable. Thorough taxonomic examination involving specimens from throughout its range would be beneficial.

top left: pistillate scales, perigynia
top right: inflorescences
bottom left and right: habit

Carex buxbaumii Wahlenb.
Common name: Brown Bog Sedge
Section: *Racemosae*   Key: F

KEY FEATURES:
- Glaucous foliage with reddish brown plant bases
- Whitish, papillose perigynia
- ♀ scales with awns

DESCRIPTION: Habit: Rhizomatous. **Culms**: 25-75 cm tall.

Leaves: 2-3.5 mm wide. Leaf sheaths ladder-fibrillose. **Inflorescences**: With 3-5 spikes, each 1-2.5 cm long, the lateral spikes ♀, the terminal spike gynecandrous. **♀ scales**: Light to dark brown with light midvein, mucronate to awned, awn if present to 3 mm long. **Perigynia**: Whitish, pale green, or gray, elliptic, strongly papillose, 2.5-4 mm long and 1.5-2 mm wide, beakless or nearly so. Stigmas 3. **Achenes**: Trigonous, nearly as big as perigynium body.

HABITAT AND DISTRIBUTION: Very discontinuous distribution in bogs and wet meadows, generally not in running water, 80-6500 feet elevation; coastal NW WA, mts of N WA, Cascades of WA and OR, mts of NE OR, and near Brookings, SW OR. AK to Greenland, south to CA and SC, also Eurasia and Algeria. Naturalized in southern Australia.

IDENTIFICATION TIPS: This distinctive species is "all field mark." Its glaucous leaves with rich reddish brown leaf bases, pale spikes with a shaggy look due to the awned, striped ♀ scales, and its pale, strongly papillose perigynia distinguish it from all similar species. *Carex livida* also has pale, papillose perigynia, but its leaves are more brightly glaucous, its terminal spike is ♂, and its ♀ scales are awnless.

COMMENTS: Rhizomes of *C. buxbaumii* have been used in basketry, like those of *C. barbarae.* The species is a community dominant in some bogs. Although it is locally common in the northern part of its circumboreal range, *C. buxbaumii* is rare in most U.S. states where it occurs, extirpated from Ireland, and rare in parts of Europe including Germany and Bulgaria. Few sedges manage to be rare in so many different jurisdictions. It is uncommon in southern Australia, where it was introduced. Historically, the main cause of its decline is conversion of land to agriculture. The main threat to today's remnant populations is habitat loss due to human activities that lower water tables, although the Cascadian populations appear secure from this threat. Additional threats to its habitats include invasive species and recreational impacts, especially by off-road vehicles and horseback riders. The tiny SW OR population has been halved by clearing for a power line and will probably soon be extirpated.

Carex buxbaumii

top left: perigynia, pistillate scale   *top right, center left*: inflorescences   *bottom*: habit

Carex californica L. H. Bailey
Common name: California Sedge
Section: *Paniceae*   Key: F

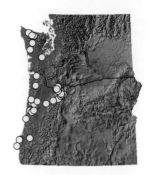

KEY FEATURES:
- Perigynia strongly papillose, with distinct beak
- Reddish brown plant bases
- Upland meadows, roadsides

DESCRIPTION: Habit: Rhizomatous. **Culms**: 15-70 cm tall. **Leaves**: Flat to folded, 10-30 (-50) cm long, 1.5-5.5 mm wide. Lower leaves reduced to bladeless sheaths. Plant bases dark reddish brown. **Inflorescences**: Lateral spikes ♀ or androgynous, 1-3.5 (-5) cm long, 2.5-7 mm wide. Terminal spike usually ♂ or androgynous. **♀ scales**: As long as the perigynia or shorter, reddish brown to purplish with lighter, papillate midrib, acute to obtuse, sometimes with a short point or awned. **Perigynia**: Reddish brown or green with reddish brown speckles or blotches, elliptic, strongly papillose, 3.3-4.3 (-5) mm long including a distinct beak 0.5-1 mm long. Stigmas 3. **Achenes**: Trigonous.

HABITAT AND DISTRIBUTION: Uncommon in meadows, open burned forests, clearcuts that are not sprayed with herbicides, and roadsides, not on serpentine substrates, W of the Cascades in WA and OR, and in the N OR Cascades. WA to CA, disjunct in ID.

IDENTIFICATION TIPS: *Carex californica* often grows mixed in other vegetation, its inflorescences the only obvious evidence of its presence, but it may form clumps on roadsides. Its perigynia are so strongly papillose that they may appear gland-covered. The rich reddish or purplish brown of its plant bases and ♀ scales distinguish it from its paler relatives. *Carex livida, C. klamathensis,* and *C. crawei* are similar, but have whitish to tan plant bases and ♀ scales, and perigynia that taper gradually to the beak. Also, *C. crawei* lacks papillae on the perigynia and is a northern species that is rare or extirpated in NE WA.

COMMENTS: *Carex californica* is an early successional plant. Potentially, clearcuts could provide good habitat, but it disappears from clearcuts that are sprayed with herbicides. It may be common where it occurs, but it has an interrupted, local distribution. Ironically, it is considered rare in California. This species is suitable for use in meadow restorations and along roadsides throughout the coast range and W Cascades, but low seed yield may limit its use. Existing plants can easily be divided to produce plugs for planting.

top left: perigynia
top right: inflorescences
center left: habit
bottom left: spike

Carex canescens L. ssp. *canescens*
Common name: Silvery Sedge
Section: *Glareosae*   Key: I

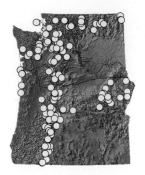

KEY FEATURES:
• Cespitose with gray-green foliage
• Perigynia small, almost beakless, usually glaucous
• Upper spikes crowded but lower ones separated

DESCRIPTION: Habit: Densely cespitose. **Culms**: 15-60 cm long. **Leaves**: (1.5-) 2-4 mm wide, usually gray-green (sometimes green), shorter than or equal to the culms. Leaf sheath fronts white-hyaline, U-shaped at summit. **Inflorescences**: Usually greenish gray to straw-colored (sometimes dark brown), 2-5 (-7) cm long with 4-8 spikes, upper spikes crowded but the lower more separated. Spikes gynecandrous, usually oblong but sometimes subglobose, 3-12 mm long with (5-) 10-20 (-30) perigynia, the beaks of the perigynia scarcely interrupting the outline of the spikes. ♀ **scales**: Usually whitish except for the midvein, to green or tinged brownish, equal to or shorter than the perigynia but not hiding them. **Perigynia**: Ascending, usually pale, gray-green, or green but in some populations yellow-brown or brown; with many obscure veins; 1.8-2.3 (-3) mm long; elliptic, with short beak. Beaks blunt, 0.3-0.5 mm long. Dorsal suture usually shorter than the beak, inconspicuous. Stigmas 2. **Achenes**: Lenticular.

HABITAT AND DISTRIBUTION: Bogs, wet meadows, wet forests, lake shores, and pond margins to 7000 feet elevation, widely distributed in mountainous and coastal areas of the PNW. AK to NF, S to CA and NC but absent from the Great Plains; also Greenland, Eurasia, southern S America, Australia, New Guinea.

IDENTIFICATION TIPS: Typical plants with glaucous foliage and whitish, almost beakless perigynia, are distinctive. Dwarf plants with short, almost round spikes and often dark perigynia can be hard to identify. *Carex praeceptorum* is similar except that all spikes are close together and overlapping, and its brown perigynia have a dark brown dorsal suture, darker than the rest of the beak. Its ♀ scales are brown at least near the midrib and the midrib is paler. It grows at high elevations (above 7200 feet in OR, 5900 feet in WA). *Carex arcta* is a greener plant with more crowded spikes and longer perigynium beaks, at least 0.75 mm long.

COMMENTS: With its tightly cespitose, grayish foliage, *C. canescens* makes an interesting garden plant in wet climates. Plants are not long-lived but readily reseed. Dwarfed plants with dark perigynia may grow with normal plants but look like a different species; is this a result of genetics, fungus, or what? Robust, dark-perigynia plants grow in mts of AK & BC, southern S America, and Falkland Islands; their relationship to typical *C. canescens* is unknown.

top left: perigynia
top right:
    inflorescence
center left:
    inflorescences
bottom: habit

Carex capillaris L.
Common name: Hair Sedge
Section: *Chlorostachyae* Key: F

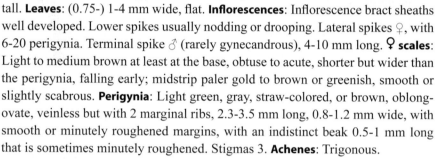

KEY FEATURES:
- Culms slender, nodding
- Spikes spreading to dangling on slender stalks
- Inconspicuous in moss mats in wetlands, usually on limestone

DESCRIPTION: Habit: Cespitose. **Culms**: (1.8-) 10-60 cm tall. **Leaves**: (0.75-) 1-4 mm wide, flat. **Inflorescences**: Inflorescence bract sheaths well developed. Lower spikes usually nodding or drooping. Lateral spikes ♀, with 6-20 perigynia. Terminal spike ♂ (rarely gynecandrous), 4-10 mm long. ♀ **scales**: Light to medium brown at least at the base, obtuse to acute, shorter but wider than the perigynia, falling early; midstrip paler gold to brown or greenish, smooth or slightly scabrous. **Perigynia**: Light green, gray, straw-colored, or brown, oblong-ovate, veinless but with 2 marginal ribs, 2.3-3.5 mm long, 0.8-1.2 mm wide, with smooth or minutely roughened margins, with an indistinct beak 0.5-1 mm long that is sometimes minutely roughened. Stigmas 3. **Achenes**: Trigonous.

HABITAT AND DISTRIBUTION: In moss mats in moist areas; bogs, marshes, stream banks, wet meadows, moist woods, up to timberline, usually but not always on limestone substrates; on flat sites to steep slopes; rare in N WA (Chelan and Okanogan cos.) and Wallowa Mts., OR. AK to NB, S to OR, NM, and ME, also Eurasia.

IDENTIFICATION TIPS: *Carex capillaris* is a delicate little sedge with dangling spikes. It might be mistaken for a grass but really isn't similar to any other sedge we have, except its alpine relative *C. tiogana*, which grows on Steens Mt. *Carex tiogana* is shorter, with falcate leaves, ♀ scales with scabrous midribs, and perigynia 1.5 – 2.1 mm long.

COMMENTS: *Carex capillaris* is rare in the PNW because habitat is scarce at the southern edge of its range. The main threats to its persistence are hydrologic change to its habitat and climate change, but recreation results in trampling of some populations. The species is variable and some botanists recognize varieties or subspecies, or even two or three similar species. Others argue that the traits do not vary together and therefore the taxa are not useful. If subtaxa are recognized within *C. capillaris*, our plants are probably *C. capillaris* var./ssp. *capillaris*.

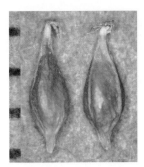

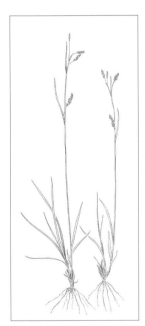

top left: perigynia
top right: inflorescence
center left: perigynia
botttom left: habit
bottom right: inflorescences

Carex capitata L.

Common name: Capitate Sedge
Section: *Capituligerae*   Key: A
Possible synonyms: *C. arctogena* Harry Smith,
   C. capitata L. ssp. *arctogena* (H. Sm.) Hiitonen,
   C. antarctogena Roivainen

KEY FEATURES:
• Single spike with spreading perigynia in lower half
• Perigynia nearly circular
• High elevations E of Cascade crest

DESCRIPTION: Habit: Cespitose or short-rhizomatous. **Culms**: 10-30 (-40) cm tall, reddish purple at the base. **Leaves**: To about 1 mm wide. **Inflorescences**: One spike per culm, 0.5-1.1 (-1.8) cm long, with spreading perigynia. ♀ **scales**: Ovate, about as long as the perigynium body, margins narrowly white-hyaline. **Perigynia**: Narrowly ovate to nearly circular, planoconvex in cross section, 2-4 mm long, 1.5-1.8 mm wide, dorsal surface with a few weak veins, ventral surface without veins, but with distinct marginal ribs; perigynia abruptly narrowing to the beak, which is 0.3-0.7 mm long; smooth or slightly serrulate. Rachilla well developed. Stigmas 2. **Achenes**: Lenticular, much smaller than perigynia.

HABITAT AND DISTRIBUTION: Wet or seasonally wet meadows, often alpine but also at lower elevations in cold air drainages or cold springs, in the PNW often on sandy, acidic soils, though worldwide it grows on diverse soils (peat, sand, gravel, often limestone), usually where snowpack is shallow but the ground remains moist in summer due to snowmelt; E of Cascades crest in OR; reported but unconfirmed in N WA. AK to Greenland, S to NH, NV, and CA, also Europe, Mexico (Chihuahua), Australia, and southern S America.

IDENTIFICATION TIPS: *Carex capitata* is an unusual species with its almost circular perigynia spreading in the lower part of its solitary spikes. *Carex nigricans* grows in similar habitat and has solitary spikes, but it is usually turf-forming and its perigynia are narrowly elliptic.

COMMENTS: *Carex capitata* can be a community dominant in moist montane to alpine meadows, and is used to retore such habitats. It survives fire well. In Scandinavia, botanists recognize two species, *C. capitata* and *C. arctogena*, distinguished by subtle morphological and habitat clues. In N America, many plants are intermediate between those two "species" and most botanists call them all one species, a variable *C. capitata*. *Carex antarctogena* of southern S America is very similar and often treated as part of *C. capitata*.

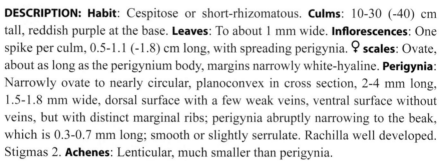

top left: perigynia
top right: inflorescences
center left and right: habit
bottom right: inflorescences

Carex chordorrhiza Ehrh. ex L. f.
Common name: Rope-root Sedge
Section: *Chordorrhizae*   Key: H, (A)

KEY FEATURES:
• Long stolons
• Bog habitat
• Compact inflorescence

DESCRIPTION: Habit: Rhizomatous but rhizomes rarely collected. **Culms**: Vegetative culms erect when young, becoming horizontal stolons 1+ m long. Flowering culms 5-35 cm tall. **Leaves**: 0.4-3 mm wide, with ligules 0.5-6 mm long. **Inflorescences**: Short, 0.5-1.6 cm long, oval, with 2-7 sessile spikes that are bractless, erect, and androgynous, or the lateral ones ♀. ♀ **scales**: Brown with greenish center and pale margins, about as long as the perigynia. **Perigynia**: Ascending, green or brown, glossy, elliptic, with many veins as well as 2 marginal ribs, 2-4 (-4.5) mm long, 1.4-2.2 mm wide. Stigmas 2. **Achenes**: Lenticular.

HABITAT AND DISTRIBUTION: Perennially wet, organic soils of bogs, sedge marshes, arctic beaches, and stream banks. Native in *Sphagnum* bogs in Okanogan Co., WA; also introduced to commercial cranberry bogs in coastal OR. AK to NB, S to OR and MT, also IA and PA; circumboreal.

IDENTIFICATION TIPS: In a well-vegetated bog, inconspicuous *C. chordorrhiza* may first be detected as a differently colored or textured patch. It is utterly distinctive once the long stolons, which root at nodes, are observed. The stolons are often covered with moss and other vegetation and appear to be rhizomes. Few *Sphagnum* bog species have a small, oval head like this one. The compact inflorescence may be misinterpreted as a single spike. *Carex simulata* is slightly similar, but does not live in acid bogs, has rhizomes rather than leafy stolons, is dioecious, has larger ♀ heads, and has dark, short-beaked, perigynia with grooved backs.

COMMENTS: *Carex chordorrhiza* is rare in the PNW, with one native WA population and two introduced populations in OR. The OR populations, accidentally introduced into cranberry bogs, threaten nearby natural bog communities. *Carex chordorrhiza* requires moderately reducing soil conditions and nearly constant shallow flooding. Threats to native populations include dewatering of bog habitats, moderate to heavy grazing, and recreational uses of habitat. Logging activities upslope of populations may cause an influx of nutrients into the bog habitat, favoring taller, competing vegetation.

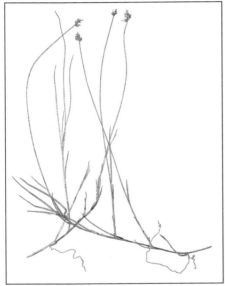

top left: perigynia
top right: inflorescence
center left: habit
middle right: stem/sheaths
bottom right: habitat (bog)

Carex circinata C. A. Mey.

Common name: Coil-leaf Sedge
Section: *Circinatae*   Key: A

KEY FEATURES:
- Single spike
- Long, narrow perigynia
- Cespitose

DESCRIPTION: Habit: Cespitose to very short-rhizomatous. **Culms**: 5-25 cm long, delicate, longer than or as long as the leaves. **Leaves**: Involute, 0.5 (-1.5) mm wide, arching to somewhat curled. **Inflorescences**: One spike per culm, (0.8-) 1.5-2.5 cm long, androgynous, lacking inflorescence bracts. ♀ **scales**: Persistent, about as long and wide as the perigynia, exposing only the beak. **Perigynia**: Ascending, linear lanceolate to narrowly fusiform, with several faint veins as well as 2 marginal ribs, 4.5-6 mm long, 0.7-0.9 mm wide, 9 times as long as wide, the distal margins finely serrulate, beak tip hyaline. Stigmas (2-) 3. **Achenes**: Trigonous (or lenticular).

HABITAT AND DISTRIBUTION: N- to NW-facing cliffs and talus slopes, also wet meadows, at 3200-4500 feet elevation, on the Olympic Peninsula, WA. In general, moist cliffs and bluffs near the coast, at low to high elevations; AK to WA.

IDENTIFICATION TIPS: *Carex circinata* is a densely tufted species with solitary spikes and the arching leaves that give it its scientific and common names. It is one of three PNW species with single spikes and very narrow perigynia. *Carex anthoxanthea* is similar but is rhizomatous with flat leaves and shorter perigynia that have smooth distal margins and dark beak tips. *Carex pauciflora* has deciduous ♀ scales, and perigynia that are widely spreading and usually longer.

COMMENTS: This species has retreated northward since the last ice age, leaving isolated remnant populations along the southern edge of its range. Specific threats to *C. circinata* populations have not been identified, but global warming is probably the greatest threat to the WA populations. It is vulnerable in WA because only two populations are known to exist there.

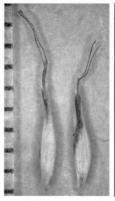

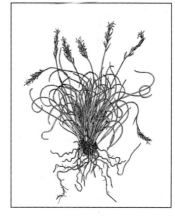

top left: habit
top middle: perigynia
top right: inflorescence
center right: habit
bottom left: inflorescences

Carex comosa Boott
Common name: Bristly Sedge
Section: *Vesicariae*   Key: E

KEY FEATURES:
- Nodding "bottlebrush" spikes with spreading, inflated perigynia
- Beak teeth long and spreading
- Cespitose

DESCRIPTION: Habit: Cespitose. **Culms**: 50-120 cm tall, usually lacking red at base. **Leaves**: 5-18 mm wide. **Inflorescences**: Lateral 2-5 spikes ♀, 2-7 cm, erect or the lower ones nodding. Terminal spike ♂, up to 6 cm, (less often with ♀ flowers also). ♀ **scales**: Lanceolate and tapering to a long, scabrous awn, shorter than the perigynia or lower ones longer. **Perigynia**: Spreading or the lower ones reflexed when mature, inflated, glabrous, with 14-22 strong veins set close together separated by less than 2 times their width, (4.8-) 6.2-8.7 mm long, 1.1-1.8 mm wide, gradually tapered to a beak 2-3.8 mm long, with teeth widely spreading, 1.3-3 mm long. Stigmas 3, style persistent, becoming curved. **Achenes**: Trigonous.

HABITAT AND DISTRIBUTION: Marshes, lake margins, wet meadows, bogs, and wet thickets; few, scattered populations in WA and OR. ON to ME, S to TX and SC, and Mexico; rare and local in a disjunct area from S BC to ID and CA.

IDENTIFICATION TIPS: *Carex comosa* is an attractive, cespitose species with large spikes. Its perigynia have long, out-curved beak teeth, and its ♀ scales have scabrous awns. *Carex hystericina* is more delicate overall and has shorter (0.3-0.9 mm) perigynium teeth that are straight to slightly divergent and perigynium veins set farther apart, separated by at least three times their width. *Carex vesicaria* and *C. exsiccata* lack spreading teeth on the perigynia and scabrous awns on the pistillate scales.

COMMENTS: *Carex comosa* is a ruderal species, establishing in disturbed wetlands and persisting. Both shoots and roots die back so thoroughly in winter that the plant can seem dead rather than dormant, but individual plants can survive for at least 20 years. In the Midwest, *C. comosa* is used for stormwater-control structures including wet meadows, rainwater gardens, and wet swales. It is not suitable for sites subject to siltation because, lacking long rhizomes, it cannot grow up out of the sediments. Commercial seed originates from the E and Midwest. The small western populations are at risk due to human-caused changes in hydrology, development of lakeshores, and competition by Reed Canarygrass. The species is rare in WA and CA, and was considered extirpated from OR until 2003, when a population was discovered in Klamath County.

top left: pistillate scales, perigynia
top right: inflorescence
center left: spike
bottom right: habit

Carex concinna R. Br.
Common name: Cute Sedge
Section: *Clandestinae*   Key: C

KEY FEATURES:
- Pubescent perigynia with crinkled hairs
- Tiny ♂ spike
- Shallow soil over calcareous substrates

DESCRIPTION: Habit: Rhizomatous, shoots arising singly or in clumps. **Culms**: 5.5-20 cm tall. **Leaves**: 5-10 cm long, 1-3.1 mm wide, basal sheaths reddish brown, fronts membranous. **Inflorescences**: Lowest inflorescence bract bladeless or with a blade less than 2 cm long, purple-tinged, with an inflated sheath 2-5 mm long. Lateral spikes 2-3, crowded near the terminal spike, 0.4-0.7 mm long, 0.2-0.4 mm wide. Terminal spike ♂, tiny, 0.3-0.7 cm long, 1.1-1.5 mm wide. ♂ **scales:** dark reddish brown with whitish margins. ♀ **scales**: reddish brown, ovate, apex obtuse, minutely ciliate, smaller than the perigynia. **Perigynia**: Pubescent with crinkled hairs, elliptic, with succulent bases that wither when dry, veinless but with 2 ribs; 2.3-3.3 (-3.5) mm long, 1.1-1.4 mm wide. Beak 0.3-0.4 mm long. Stigmas 3. **Achenes**: Rounded to trigonous. **Anthers:** 1.3-1.4 mm long.

HABITAT AND DISTRIBUTION: Streambanks, bogs, meadows, and open forest, on stable, usually calcareous substrates that are wet in spring but may dry out later. NE OR. AK to S BC (Selkirk Mts), to NF, S to OR, CO, SD, and MI.

IDENTIFICATION TIPS: *Carex concinna* has hairy perigynia, a tiny staminate spike, and no inflorescence bract (or a tiny one). *Carex concinnoides* is similar but has 4 stigmas and a larger ♂ spike. *Carex rossii* and *C. deflexa* have basal spikes among the leaf sheaths, longer inflorescence bracts, larger ♂ spikes, and fine, straight perigynium hairs. *Carex rossii* has a longer perigynium beak. *Carex inops* grows in the Cascades and has a longer ♂ spike held above the ♀ spikes.

COMMENTS: *Carex concinna* is a northern species that retreated northward as the glaciers melted, leaving behind isolated populations in mountains and around the Great Lakes. It is confined to shallow soil over rock, streambanks, and other areas unsuitable to tall, competing vegetation. On lakeshores in the NE U.S. it is threatened by recreational foot traffic and off-road vehicles. Trampling by livestock threatens some populations. *Carex concinna* appears more secure in the PNW, but its limited range and specific habitat requirements make it vulnerable. The tiny ♂ spike and small anthers suggest that *C. concinna* may be largely self-pollinating.

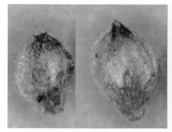

top left: perigynia
top right: inflorescences
center left: perigynia; habit
bottom right: habit

Carex concinnoides Mack.
Common name: Northwestern Sedge
Section: *Clandestinae*   Key: C

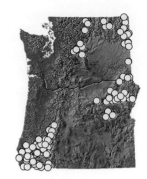

KEY FEATURES:
- Pubescent perigynia with four stigmas
- Blooms very early in spring
- Dry woods and openings

DESCRIPTION: Habit: Rhizomatous, the shoots usually arising singly and spaced well apart, sometimes a few together. **Culms**: 14-37 cm tall, arching to the ground as the perigynia mature. **Leaves**: 6-25 cm long, 1.7-4.5 mm wide, basal leaf sheaths reddish brown, fronts membranous, mouth with minute cilia. **Inflorescences**: 1.2-3.3 cm long, bract reddish-brown, bristle-like, sheathless, 5-8 mm long, sometimes with an awn to 1 (-6) mm long. Lateral spikes 1-3, crowded near the terminal spike, 0.6-1.3 cm long, 3-6 mm wide. Terminal spike ♂, 0.8-2.2 cm long, 1.8-3.1 mm wide. **♂ scales:** Dark reddish brown or purplish black, with whitish margins. **♀ scales**: Dark reddish brown, ovate or obovate, apex obtuse to acute, minutely ciliate, smaller than the perigynia. **Perigynia**: Pubescent with straight hairs, elliptic to obovoid, with succulent bases that wither when dry, veinless but with 2 ribs, 2.5-3 mm long, 1.4-1.7 mm wide. Beak 0.5 mm long. Stigmas 4. **Achenes**: Rounded but with somewhat quadrangular bases. **Anthers:** 1.9-3.1 mm long.

HABITAT AND DISTRIBUTION: Open conifer forests and sometimes on serpentine substrates, at low to moderate elevations in mts; mts of WA E of the Cascade crest, NE OR, and SW OR N to W Cascades of Lane Co. BC to SK, south to CA and MT.

IDENTIFICATION TIPS: *Carex concinnoides* is the only N American sedge with 4 stigmas. It blooms early in spring, before nearly all other flowers. Its loosely spaced shoots or tufts with dark green, arching leaves make even sterile plants identifiable. *Carex concinna* differs in having a tiny ♂ spike. *Carex rossii* and *C. deflexa* differ in having basal ♀ spikes in addition to the ones near the ♂ spike, a more condensed growth form, narrower leaves, and less arching leaves and culms. In SW OR, *C. serpenticola* grows in the same habitat and flowers at the same time, but it has 3 stigmas and is usually dioecious.

COMMENTS: *Carex concinnoides* may be an understory dominant in open forests E of the Cascade crest. Shallow rhizomes make *C. concinnoides* vulnerable to fire, and also accessible for use by humans who gathered the plants to line cooking pits and for other household uses. The long, pale yellow anthers and the many white stigmas contrast strikingly with the dark scales, attracting insects that presumably pollinate the flowers.

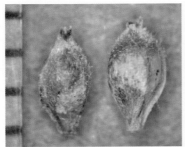

top left: perigynia
top right: inflorescences
center left: inflorescence in
    flower
bottom right: habit

Carex cordillerana Saarela & B. A. Ford

Common name: Cordilleran Sedge
Section: *Phyllostachyae*   Key: F, (A)
Synonyms: *C. backii* Boott, in part

KEY FEATURES:

- Long, leaf-like ♀ scales
- Soft, broad leaves
- Growing in shade of deciduous trees or shrubs

DESCRIPTION: Habit: Cespitose. **Culms**: 5-40 cm tall. **Leaves**: 1.5-5.9 mm wide, longer than the inflorescences. **Inflorescences**: Some inflorescences may be on relatively long culms, but others nestle among the basal leaf sheaths; bract of lowest spike leaf-like or absent. Lateral spikes 0-3, with a total of 3-5 perigynia. Terminal spike androgynous, the ♂ portion forming a tiny cylinder 1.7-2.6 mm long, hidden by the perigynia. **♀ scales**: Scales subtending the lower perigynia green, leaf-like, 2.5-3 mm wide, often 1-2 (-7) cm long, longer than the perigynia and hiding them. Scales of upper perigynia ovate, shorter than the perigynia. **Perigynia**: Green, obovoid, 3.9-5.4 mm long, 1.6-2.5 mm wide, abruptly tapered to a beak 0.5-1.6 mm long. Stigmas 3. **Achenes**: Rounded-trigonous and tightly enveloped by the perigynium.

HABITAT AND DISTRIBUTION: Rocky slopes, in leaf litter and duff, usually in shade of deciduous trees and shrubs, occasionally juniper; mts of E WA and E OR. BC to AB, south to OR, UT, and WY.

IDENTIFICATION TIPS: *Carex cordillerana* is a tufted, grass-like plant, relatively wide-leaved considering its habitat. In OR and WA, it is distinctive, once the unusual inflorescence has been found and the leaf-like lower ♀ scales have been interpreted correctly. The very similar *C. backii,* occurring in BC, has perigynia that more loosely enclose the achene and are gradually tapered to a longer beak.

COMMENTS: *Carex cordillerana,* recently split from *C. backii,* is the rare PNW representative of a complex of sedges occurring in upland forests. Populations are few and small. It gives the impression of a forest species barely holding on in difficult habitat. It always grows in the shade of Mountain Mahogany (*Cercocarpus ledifolius*), cherry (*Prunus* sp.), juniper (*Juniperus* sp.), or other trees, where it receives bright indirect light. It seems to require leaf litter and duff. Its broad, palatable leaves attract grazers, which can easily pull up the shallow-rooted plant. Perhaps it was once more widespread and now persists only on rocky slopes where it receives some protection from livestock.

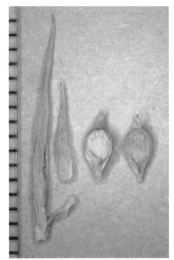

top left: dried pistillate scales and perigynia
top middle: fresh perigynia
top right: inflorescence
bottom: habit

Carex crawei Dewey
Common name: Crawe's Sedge
Section: *Granulares*   Key: F

KEY FEATURES:
- Perigynia elliptic, light green to tan, often with reddish speckles
- Rhizomatous, the shoots single
- Shallow, seasonally wet soils over limestone bedrock

DESCRIPTION: Habit: Rhizomatous, the shoots arising singly or rarely a few together. **Culms**: 2-30 (-40) cm tall, plant bases dull brown. **Leaves**: Light green, blades tending to curve back, short (0.6-9.7 cm long), 1.5-4 mm wide. Lower leaves with blades. Plant bases greenish or dull brown. **Inflorescences**: Lowest inflorescence bract with inflated sheath 5-15 mm long. Lateral 2-4 (-6) spikes ♀, (0.5-) 1-2.5 cm long, 3.1-6.3 mm wide, with 15-50 perigynia, the lowest usually arising from the lower half of the culm. Terminal spike ♂, (0.5-) 1-2.5 cm long, usually held 1-9 cm above the uppermost ♀ spike. **♀ scales**: Shorter and narrower than the perigynia, brown with paler midrib, often short-awned. **Perigynia**: Yellow-green to pale brown, often with reddish brown speckles, elliptic to ovate-elliptic, 2.2-3.4 (-3.7) mm long, 1.1-1.9 mm wide, somewhat inflated but flattened on the ventral side, with 15-25 veins. Beak 0.1-0.4 mm long. Stigmas 3. **Achenes**: Trigonous.

HABITAT AND DISTRIBUTION: Moist meadows, fens, swales, and ditches, in vernally moist areas that are often very dry in late summer, generally on limestone or other calcareous bedrock; near Colville, WA. S BC to NF, south to N WA, UT, and GA.

IDENTIFICATION TIPS: *Carex crawei* is an inconspicuous sedge with ♀ spikes low on the culm as well as up near the single ♂ spike. It is somewhat like *C. livida,* but that species has strongly glaucous leaves, spikes only near the top of the culm, and papillate, unspotted perigynia.

COMMENTS: *Carex crawei* has a very patchy distribution and is a listed rare plant in most jurisdictions in which it occurs. It may depend on disturbances caused by fire or flooding to create open habitats, but both succession to forest communities and extensive habitat disturbance threaten populations. Around the Great Lakes, it is a community dominant with Tufted Hairgrass (*Deschampsia cespitosa*) in shallow, vernally wet soil over limestone bedrock. The one WA specimen, now at the Smithsonian, was collected before 1850 at Ft. Colville, in what is now WA. Conversion of riparian zones to agriculture and dams that prevent flooding may have contributed to the extirpation of what may always have been a rare plant in the PNW.

138

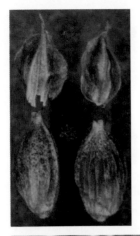

top left: perigynia, pistillate scales
top right: perigynia, habit, inflorescence
bottom left: inflorescences

Carex crawfordii Fernald
Common name: Crawford's Sedge
Section: *Ovales* Key: J

KEY FEATURES:
- Cespitose, with gynecandrous spikes and winged perigynia
- Perigynia narrowly lanceolate
- Habitat vernally wet pond margins

DESCRIPTION: Habit: Cespitose. **Culms**: 25-60 (-85) cm tall. **Leaves**: 7-22 cm long, 2-4 mm wide. **Inflorescences**: 1.8-3 cm long, 8-14 mm wide, the spikes crowded into one dense head, or the lower somewhat remote. Spikes gynecandrous. ♀ **scales**: Gold to dark brown with midstripe lighter or darker, lanceolate, 3-3.8 mm, shorter and narrower than the perigynia. **Perigynia**: Ascending, whitish to light brown, lanceolate (often narrowly so), flat except over the achene or sometimes planoconvex, 3.4-4.1 (-4.7) mm long, 0.9-1.3 mm wide, veinless or with up to 5 veins on each face, wing 0.1-0.2 mm wide. Beak winged and more or less ciliate-serrulate almost to the tip, (1.8) 2.1-3 mm from achene top to beak tip. Stigmas 2. **Achenes**: Lenticular, 1.1-1.5 mm long, 0.6-0.8 mm wide, 0.15-0.35 mm thick.

HABITAT AND DISTRIBUTION: Pond margins that are wet in spring and other seasonally wet spots; native populations in mts of N WA, near Crater Lake National Park, and in the Rogue-Umpqua Wilderness in OR; introduced to cranberry bogs and road ditches in coastal WA and OR. AK to NF, south to OR, ID, MN, and MA, introduced to MO.

IDENTIFICATION TIPS: *Carex crawfordii* is a tufted, yellow-green plant with a tight head, that grows in the "bathtub ring" around a seasonal pond. It resembles *C. athrostachya* but the inflorescence bracts are usually not prolonged beyond the end of the inflorescence, and the spikes and heads are more slender.

COMMENTS: This species raises the interesting question of how to give legal protection to a very rare native plant when it has also been introduced from a non-local source and is a minor weed in part of its PNW range. *Carex crawfordii* can bloom in its first year and function as a facultative annual.

top left: perigynia
top right: inflorescence
center left:
    inflorescences
bottom right: habit

Carex cusickii Mack. ex Piper & Beattie
Common name: Cusick's Sedge
Section: *Heleoglochin*   Key: H

KEY FEATURES:
- Leaf sheath fronts red-dotted and coppery colored near the top
- Inflorescence wider near base, branched, partly drooping, rough
- Forming large tussocks

DESCRIPTION: Habit: Cespitose, well developed plants forming tussocks 0.3 m or more in diameter at the base. **Culms**: (5-) 30-200 cm tall, plant bases blackish. **Leaves**: 2.5-5 (-6) mm wide. Leaf sheath fronts coppery-tinged near the mouth and red-dotted, sometimes cross-corrugated. **Inflorescences**: (2-) 3-8 cm long, 10-20 mm wide, branched, nodding, somewhat interrupted. Spikes androgynous or ♀; the inflorescence usually bisexual but sometimes all ♀. **♀ scales**: About as wide and long as the perigynia, usually hiding them. **Perigynia**: Light to dark brown, shiny, planoconvex (or with the ventral surface slightly convex), lance-ovate, the dorsal surface with a shallow lengthwise groove and generally also with 7-11 veins, 2.4-3 mm long, 1.3-1.8 wide. Perigynium bases swollen with pithy tissue. Beak 1-1.3 mm long. Stigmas 2. **Achenes**: Lenticular.

HABITAT AND DISTRIBUTION: Still and slow-moving waters, including bogs, fens, marshes, lakeshores, wet meadows, floating logs, and road ditches; in and W of the Cascades, also mts of N WA and E OR. BC to WY and CA.

IDENTIFICATION TIPS: *Carex cusickii* has a rough, branched inflorescence and shiny, dark, dorsally grooved perigynia. Well developed plants form large tussocks that are utterly distinctive. Smaller plants that are merely cespitose can be confused with *C. diandra,* which has whitish hyaline leaf sheath fronts, erect inflorescences, and perigynia that are not hidden by the ♀ scales.

COMMENTS: *Carex cusickii* is widespread in the PNW, but lacking from many habitats that appear suitable. This species can be a community dominant, often with *Sphagnum* moss or Bog Buckbean (*Menyanthes trifoliata),* but it may grow as a few isolated plants. We have collected culms 2 meters long from tussocks that were 0.5 meter across. The large tussocks can serve as substrate for growth of other plants. *Carex cusickii* is tolerant of serpentine soils.

top left: perigynia
top right: inflorescences
center left: inflorescence
center right: leaf sheath mouths
bottom: habit, habitat

Carex davyi Mack

Common name: Davy's Sedge
Section: *Ovales*   Key: J
Synonyms: *Carex constanceana* Stacey

KEY FEATURES:
- Cespitose, with gynecandrous spikes and winged perigynia
- Large perigynia
- ♀ scales revealing more than half the perigynium beak

DESCRIPTION: Habit: Cespitose. **Culms**: 25-45 cm. **Leaves**: 4-15 cm long, 2-3 mm wide. **Inflorescences**: 2.5-4 cm long, 13-15 mm wide, erect, the approximately 3(-7) spikes distinct. Spikes gynecandrous. **♀ scales**: Reddish brown or coppery with lighter midstripe, lacking white margins or with white margins up to 0.2 (occasionally to 0.4) mm wide, the scales 4.5-5.8 mm long, shorter and narrower than the perigynia, revealing most of the beak. **Perigynia**: Appressed to ascending, greenish to straw-colored, lanceolate to ovate, 6-8.5 mm long, (1.5-)2-2.4 mm wide, with 8-16 dorsal veins and 4-7 ventral veins, 3 or more of them extending beyond the achene onto the beak. Beak tip unwinged, brown, parallel-sided, and entire for the distal 0.5-0.9 mm, (2.6-)3-4.2 mm from top of achene to beak tip, dorsal suture inconspicuous, occasionally winged nearly to the tip. Stigmas 2. **Achenes**: Lenticular, 2-3 mm long, 1.1-1.5 mm wide, 0.4-0.7 mm thick.

HABITAT AND DISTRIBUTION: Seasonally moist scablands, open conifer forest, and ephemeral watercourses; Lake Co., OR, Mt Adams, WA (possibly extirpated), and Sierra Nevada and NE CA.

IDENTIFICATION TIPS: *Carex davyi* is a medium-sized, cespitose sedge with large perigynia and ♀ scales that reveal most of the perigynium beak. *Carex petasata,* a widespread species of sagebrush steppe, has similar large perigynia, but its ♀ scales are nearly as long as the perigynia, concealing most of the beak, and have wide white margins (0.2-0.7 mm wide). *Carex praticola* has shorter perigynia and ♀ scales with mostly wider (0.1-0.3 mm wide) white margins.

COMMENTS: This species was collected on Mount Adams, WA, in 1909 by important PNW botanist Wilhelm Suksdorf, and called *C. constanceana*. It was never found there again and was presumed extinct until 1996 when it was collected in Lake County, OR. Subsequent genetic and morphological research has demonstrated that *C. constanceana* is the same species as *C. davyi*, which had been considered endemic to the northern Sierra Nevada in California.

Carex davyi

top left: pistillate scales, perigynia
top right: inflorescence
bottom left: habit

Carex deflexa Hornem. var. *boottii* L. H. Bailey

Common name: Mountain Mat Sedge
Section: *Acrocystis*   Key: C
Synonyms: *C. brevipes* W. Boott ex Mackenzie,
   C. rossii Boott var. *brevipes* (W. Boott ex Mackenzie)
   Kükenthal, often included in *C. rossii* without
   recognizing varieties.

KEY FEATURES:
- Pubescent, short-beaked perigynia
- Basal ♀ spikes
- Forming mats in mountains, mostly above 3000 feet

DESCRIPTION: Habit: Cespitose to mat-forming. **Culms**: 5-31 cm tall. Plant bases reddish to purplish or dark brown, often with persistent leaves of previous years. **Leaves**: Green, usually longer than the culms, sometimes shorter, 0.9-2.6 (-3.2) mm wide, not densely papillose on the lower surface. **Inflorescences**: Bract of lowest lateral spike leaf-like, longer than the inflorescence. Lateral spikes (just below terminal spike) 1-4, ♀, each 0.3-0.8 cm long, with 4-15 perigynia. Terminal spike ♂, 0.3-1.1 cm long. There are also 0-2 basal ♀ spikes among the leaf sheaths at the base of each culm. ♀ **scales**: Pale to dark reddish brown with white margins, shorter than the perigynia, apex acute to acuminate. **Perigynia**: Pubescent, elliptic to obovoid, with succulent bases that wither when dry, veinless except for the 2 ribs; perigynia of lateral spikes 2.3-3.1 mm long, 1-1.4 mm wide; beaks 0.4-0.8 mm long, with apical teeth 0.1-0.2 mm long. Stigmas 3. **Achenes**: Trigonous.

HABITAT AND DISTRIBUTION: Establishing in disturbed openings and persisting in open forest, at elevations where there is snow for much of the winter, generally above 3000 feet; mts of WA and OR. BC to SK, S to CA and CO.

IDENTIFICATION TIPS: *Carex deflexa* var. *boottii* is a short, high elevation sedge with pubescent perigynia, some of them on basal spikes nestled among the leaf sheaths. In the field, it is finer-textured and more mat-like than *C. rossii*. *Carex rossii* typically has a longer perigynium beak (0.8-1.6 mm) and tougher leaves. *Carex inops* ssp. *inops* has a much more open, rhizomatous growth form, inflorescence bracts that are shorter than the inflorescence and not leaf-like, and no basal ♀ spikes.

COMMENTS: The small upland sedges with basal spikes do not fall into two neat groups corresponding to *C. deflexa* and *C. rossii*. There is more overlap in beak lengths than is reported in the literature. These sedges have enough variation in habitat and leaf morphology for three or more species, but how to distinguish the taxa (if they exist) is not at all clear.

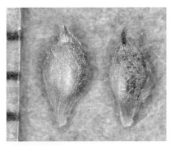

top left: perigynia
top right: inflorescences
center left: habit
bottom right: habit
bottom left: inflorescences

Carex densa (L. H. Bailey) L. H. Bailey

Common name: Dense Sedge
Section: *Multiflorae*   Key: H
Synonyms: *C. breviligulata* Mackenzie, *C. dudleyi* Mackenzie, *C. vicaria* L. H. Bailey, *C. vulpinoidea* Michx. var. *vicaria* (L. H. Bailey) Kükenthal

KEY FEATURES:
• Cross-corrugated leaf sheath fronts
• Inflorescences broadly cylindric, dense, uninterrupted
• Seasonal wetlands W of the Cascades

DESCRIPTION: Habit: Cespitose. **Culms**: 50-110 cm tall. **Leaves**: 5 mm wide, generally shorter than the flowering culm. Leaf sheath fronts cross-corrugated, white-hyaline with pale brown spots. **Inflorescences**: 3-5.7 cm long, 15-25 mm wide, dense and cylindrical, with relatively inconspicuous bracts. Spikes androgynous. ♀ **scales**: Brown, short-awned. **Perigynia**: Yellowish brown to brown, reddish brown distally, the body elliptic to ovate, 2.8-4 mm long, 1.5-2 mm wide, with (3-) 5-7 dorsal veins and (0-) 3-5 ventral veins. Base usually somewhat swollen with spongy tissue and rounded, or sometimes lacking such tissue and tapered. Stigmas 2. **Achenes**: Lenticular.

HABITAT AND DISTRIBUTION: Moist areas that may dry out in summer, such as wet meadows, wet prairies, ditches, and hillside drainages; SW WA to CA, W of the Cascades and Sierra Nevada. Introduced to Australia.

IDENTIFICATION TIPS: *Carex densa* is cespitose and has ragged, yellowish green or yellowish brown inflorescences. The superficially similar *Ovales* (which have gynecandrous spikes) and *C. hoodii* lack *C. densa's* cross-corrugated leaf sheath fronts. Closely related *Carex vulpinoidea,* found mainly E of the Cascades, is very similar but has a longer, interrupted inflorescence with a conspicuous, hairlike bract at the base of each spike, and browner perigynia that average smaller (2-3.5 mm long) with 0-3 dorsal veins and usually no ventral veins.

COMMENTS: *Carex densa* can be a community dominant in wet prairie W of the Cascades, with Tufted Hairgrass (*Deschampsia cespitosa*) or *C. unilateralis.* This plant community has been nearly eliminated by agricultural development. *Carex densa* remains common in OR, though very rare in WA. It is used in habitat restoration projects and plantings for erosion control and stormwater retention. *Carex densa* is highly variable; up to four species have been recognized within it. Venation and the development of pithy tissue vary continuously, and the variants should not be given species names even though individuals that completely lack pithy tissue have odd-looking, almost diamond-shaped perigynia.

148

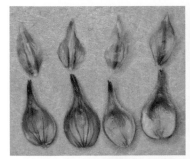

top left: pistillate scales, perigynia
top right: inflorescence
center left: leaf sheath front
bottom: habit, habitat

Carex deweyana Schwein. var. *deweyana*
Common name: Dewey's Sedge
Section: *Deweyanae*　　　Key: I

KEY FEATURES:
- Loosely cespitose, with lax, grass-like leaves
- Moist forest
- Thin, membranous perigynia, with short beaks and teeth
- Ligules about as long as wide

DESCRIPTION: Habit: Cespitose. **Culms**: 14-90 (100) cm long, smooth or papillose (at 20X) at mid-length, the fragile papillae like tiny pegs projecting at right angles from the edge of the culm. **Leaves**: 0.6-4.2 mm wide, the ligule of uppermost culm leaf rounded, only about as long as wide (0.9-2.2 mm long). **Inflorescences**: Spikes 2-5, green and white, usually gynecandrous (or ♀), lowest spikes 5.2-11 (-13) mm long, with (5-) 7-13 perigynia; terminal spike 6.8-13 mm long. ♀ **scales**: White (to straw-colored) with green midvein, usually awned, the awn to 0.3 mm. **Perigynia**: 4-4.9 (-5.2) mm long, with beak 1.4-2.1 mm long, 30-40% of perigynium length, with teeth to 0.2 mm long. Stigmas 2. **Achenes**: Lenticular. **Anthers:** (1.8-) 1.9-2.2 mm long.

HABITAT AND DISTRIBUTION: Shaded to partially shaded riparian zones and moist forest; mts of NE WA. AK to NF, south to N WA, CO, ND, and IL.

IDENTIFICATION TIPS: Like its more widespread PNW relatives *C. bolanderi, C. infirminervia*, and *C. leptopoda, C. deweyana* var. *deweyana* is loosely cespitose with green, lax, grass-like leaves, separated spikes, and slender perigynia. It is distinguished by its short spikes with few perigynia that are relatively long, and short ligules. The ligule is short both because the culm is narrow and because the blade joins the sheath at a nearly horizontal angle. The ligule should be measured from where it meets the leaf midrib to where it meets the leaf edge. This taxon can be confused with depauperate *C. leptopoda,* which has smaller perigynia.

COMMENTS: Until recently, *C. deweyana* was considered very common in moist forest but now it has been split into four PNW species. *Carex deweyana* as now understood is a northern species occurring in the PNW only in mountains of NE WA. See also *C. bolanderi, C. infirminervia,* and *C. leptopoda.*

Carex deweyana var. *deweyana*

top left: fresh perigynia
top center: dried
    perigynium
top right: inflorescences
center: ligule
bottom: habit

Carex diandra Schrank
Common name: Lesser Panicled Sedge
Section: *Heleoglochin*   Key: H

KEY FEATURES:
- Shiny, dark perigynia with dorsal groove and bulging ventral surface
- Leaf sheath fronts white with red dots
- Bog habitat

DESCRIPTION: Habit: Cespitose to short-rhizomatous. **Culms**: 2-90 cm tall. **Leaves**: 1-2.5 mm wide. Leaf sheath fronts white (or sometimes coppery tinged) with red dots. **Inflorescences**: 1.7-4.0 (-6.0) cm long, 6-15 mm wide, inconspicuously branched, erect, not interrupted or slightly so. Spikes androgynous or ♀. ♀ **scales**: About as wide as the perigynia at the base, but shorter than the perigynia and narrowing toward the tip and thus not concealing the perigynia. **Perigynia**: Olive brown to dark brown and sometimes green on beak, shiny, biconvex, lance-ovate, so filled with pithy tissue that the ventral surface often bulges, the dorsal surface with a shallow narrow groove and 4-6 veins. Beak 0.9-1.1 mm long. Stigmas 2. **Achenes**: Lenticular.

HABITAT AND DISTRIBUTION: Bogs and fens, floating peat mats, lakeshores, springs, and seeps; in mts of N WA, Cascades of WA and OR, and in Lake Co., OR. AK to NF, south to CA and MD, Iceland, Europe, Canary Islands, Asia S to the Himalayas and New Zealand.

IDENTIFICATION TIPS: *Carex diandra* is an inconspicuous bog plant with red-dotted leaf sheath fronts, brown inflorescences, and perigynia that are shiny, dark, dorsally grooved, and ventrally bulging. It is easily confused with small individuals of *C. cusickii,* which has conspicuous coppery coloration at the top of the leaf sheath front. Typically *C. cusickii* differs in its larger, obviously branched inflorescences and its tendency to form large tussocks. *Carex vulpinoidea* differs in having strongly cross-corrugated leaf sheath fronts, awned ♀ scales, and usually longer inflorescences.

COMMENTS: Worldwide, *C. diandra* inhabits diverse wetlands with relatively stable water levels dependent on groundwater rather than surface runoff, usually on peat soils. It inhabits the wettest nonaquatic microsites in diverse fens with pH usually between 6 and 8. *Carex diandra* is circumboreal and common N of the lower 48 U.S. states. In the continental U.S., populations are few and mostly isolated, and the species is listed as rare in at least 10 states. The main threats to its existence are hydrologic change and global warming. Off-road vehicle use threatens some populations.

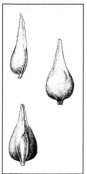

top left: perigynia   *top middle*: habit   *top right*: inflorescence
center left: perigynia
bottom: habitat

Carex disperma Dewey
Common name: Two-seed Sedge
Section: *Dispermae*　　　Key: H

KEY FEATURES:
- Spikes with 1-3 perigynia, scattered on very delicate culms
- Perigynia egg-shaped, becoming dark brown
- Shady, moist habitats

DESCRIPTION: Habit: Rhizomatous, with slender brown rhizomes. **Culms**: 15-60 cm long, longer than the leaves, very slender, erect when young but usually arching when mature. **Leaves**: Green, flat, 0.7-1.5 mm wide. **Inflorescences**: 1.5-3.5 cm long, with lowest inflorescence bract 5-20 mm long, with 1-6 androgynous spikes, each with only 1-2 inconspicuous ♂ flowers in addition to 1-3 (-6) perigynia, the lower spikes separated but the uppermost crowded together. ♀ **scales**: White hyaline with green midrib, shorter and narrower than the perigynia. **Perigynia**: Plump ovate, green when young but becoming dark brown, shiny, veined, with short beak only about 0.25 mm long. Stigmas 2. **Achenes**: Lenticular.

HABITAT AND DISTRIBUTION: Moist forests, aspen groves, bogs, swamps, very often on small hummocks in forested wetlands, often growing in a moss layer, usually in shade, sea level to 10000 feet elevation; widespread in mts, in and E of the Cascades, and in the Olympic Mts., WA. AK to NF, S to CA, NM, and PA, also Eurasia.

IDENTIFICATION TIPS: The problem with this delicate sedge is seeing it, not identifying it once it's found. The almost thread-like culms with widely separated clusters of egg-shaped, brown perigynia are unique. *Carex brunnescens* has thicker culms and more perigynia per spike. *Carex laeviculmis* could be considered vaguely similar but has thicker culms and its spikes have 5-15 beaked perigynia.

COMMENTS: Because *Carex disperma* is so hard to see, particularly when growing in a stand of other herbaceous plants, it is easy to overlook. It is vulnerable to hydrologic changes, to logging that removes the shade it seems to need, and to trampling.

top left: perigynia
top right: inflorescences
center left: inflorescence
bottom right: habit

Carex distans L.
Common name: Distant Sedge
Section: *Spirostachyae*   Key: E, F

KEY FEATURES:
- Cespitose
- Spikes distant from one another
- Perigynia glabrous, with distinct beak

DESCRIPTION: Habit: Cespitose. **Culms:** 40-80(100) cm tall, bases brown to orange-brown. **Leaves:** 3-4.6 mm wide, shorter than the culms. **Inflorescences:** Lowest inflorescence bract leaf-like, much shorter than inflorescence, with sheath >4 mm long; lateral 2-4 spikes ♀, ascending; terminal spike ♂, (15-)20-32(-40) mm long, on stalk 2-5(-8.5) cm long. ♀ **scales**: Brown, with midrib green or whitish, 3-veined. **Perigynia:** Elliptical, gray green to greenish brown, slightly inflated near the top, 3.5-4.6 x 1.1-1.4 mm, with 10-16 veins, contracted to scabrous beak (0.8-)1-1.4 mm long. Styles deciduous but leaving a relatively long, often contorted stub on the achene. Stigmas 3. **Achenes:** Trigonous.

HABITAT AND DISTRIBUTION: Freshwater and brackish wetlands, wet meadows, depressions in sand dunes; near the Columbia River in Morrow and Umatilla counties, OR; central and southern Europe, North Africa, southeast Asia; introduced to PA and MA.

IDENTIFICATION TIPS: *Carex distans* is a conspicuously cespitose sedge with a single staminate spike well separated from the pistillate spikes, which are in turn separated from each other. Perigynia are trigonous and distinctly beaked.

COMMENTS: *Carex distans* was found in or near ballast dumps in PA (1865) and MA (1948), but probably does not persist there. It was collected in the Irrigon State Wildlife Refuge in 2010, among sand dunes near the Columbia River. It seems well established there with a population of hundreds in an alkaline wetland with sandy soil. It should be looked for in wetlands along the lower Columbia River. In its native range, this species is common near the coast and in alkaline desert wetlands, though it can occur elsewhere. In Europe, *C. distans* hybridizes with several other species including *C. viridula*, a wide-ranging species which is native in Oregon.

Inland sand dunes fluctuate in response to climate change and human activities. Hydroelectric dams and conversion to irrigated agriculture have reduced dunes in Washington to a quarter the area they occupied before European settlement. Abundant invasive plants such as Cheatgrass (*Bromus tectorum*) threaten remaining dunes. New invaders such as *C. distans* may cause further impacts as they establish and spread.

top left: perigynia
top center and right:
 inflorescences
center left: plant
 bases
bottom: habitat
 (stabilized
 inland dunes)

Carex divulsa Stokes
Common name: Grassland Sedge
Section: Phaestoglochin   Key H, J

KEY FEATURES:
- Cespitose
- Inflorescence linear, interrupted
- Perigynia planoconvex, slightly winged

DESCRIPTION: Habit: Cespitose. **Culms**: 25-90 cm tall. **Leaves**: Widest 2-3 mm wide, often papillose on upper surface. **Inflorescences**: 5-18 cm long with 4-8 spikes, interrupted, the lowest internode usually at least 2 times longer than the lowest spike or side branch; lowest node usually with 2+ spikes. Spikes gynecandrous. ♀ **scales**: 2.8-3.7(-4) mm long, hyaline or pale brown with green, acute or short-awned. **Perigynia**: 3.5-5.5 mm long, 2-2.6 mm wide, spreading, somewhat winged on the beak, pale yellow or tan becoming dark brown to black, usually with up to 11 dorsal veins; beak 0.8-1.5 mm long. Stigmas 2. **Achenes**: Lenticular, 2.3-2.5 mm long x 1.5-1.7 mm wide.

HABITAT AND DISTRIBUTION: Fields, hedgerows, open forest at low elevations; introduced in NW WA and CA, native to Europe and W Asia, naturalized Australia.

IDENTIFICATION TIPS: *Carex divulsa* is a cespitose sedge with a long, interrupted inflorescence with short spikes. It somewhat resembles *C. vallicola* and *C. vulpinoidea* but has a more interrupted inflorescence and larger perigynia. *Carex divulsa* is often sold as a lawn cover under the name of *C. tumulicola*, a native species that has slightly narrower leaves, shorter, ascending perigynia, and usually fewer spikes. The lowest inflorescence internodes of *C. tumulicola* are shorter than to less than 2 times as long as the lowest spike.

COMMENTS: *Carex divulsa* is sold as a low-maintenance, tough, evergreen, grass-like plant for lawns, meadows, and edges of trails. It is tolerant of shade, wet soils, and trampling, and is drought-tolerant once established. It is also a prolific seeder that invades native habitats. It has become a dominant ground cover in some California coastal forests.

Two subspecies of *C. divulsa* have been reported in the PNW. Some plants appear intermediate. In its native range, *C. d.* ssp. *leersii* is restricted to limestone and chalk substrates. The subspecies can be separated as follows:

1a. Leaves gray-green to dark green; inflorescence 5-18 cm long with lower
    spikes >2 cm apart; perigynia 3-4(-4.5) mm long, usually greenish-brown
    when ripe..*C. d.* ssp. *divulsa*
1b. Leaves usually yellowish-green; inflorescence 4-8 cm long with lower spikes
    usually <2 cm apart; perigynia 4-4.5(-4.8) mm long, usually reddish-brown
    when ripe.. *C. d.* ssp. *leersii* (Kneuck.) W. Koch

top: inflorescences
center left: perigynia
center right: lateral spike
bottom: habit

Carex douglasii Boott in W. J. Hook.
Common name: Douglas' Sedge
Section: *Divisae*   Key: B, H

KEY FEATURES:
- Well-spaced, short shoots on brown, thin rhizomes
- Separate ♂ and ♀ plants, the ♀ heads large and shaggy
- Seasonally moist spots in sage steppe

DESCRIPTION: Habit: Rhizomatous, the rhizomes brown, slender, 0.8-1.9 mm in diameter. **Culms**: (8-) 15-40 cm. Shoot bases brown. **Leaves**: 1-3 (-3.5) mm wide. **Inflorescences**: 1.2-3.5 (-4.5) cm long, ♂ inflorescenses 7-15 mm wide, ♀ inflorescences 8-27 mm wide. Plants apparently dioecious; flowers in one plant seemingly all ♀ or all ♂ but usually with a few, nearly undetectable flowers of the opposite sex mixed in the spikes. ♀ **scales**: Pale brown to whitish hyaline, 4.3-4.7 (-7.5) mm, awned or not. **Perigynia**: Brown, variable between plants though consistent within one plant, ovate to obovate, (3-) 3.5-4.2 (-4.8) mm long, 1.2-2.1 mm wide. Beak (0.9-) 1.2-1.9 mm long. Stigmas 2, long, tangled, often retained at perigynium maturity. **Achenes**: Lenticular. **Anthers:** Tipped with a tiny, bristly awn, often retained long past flowering.

HABITAT AND DISTRIBUTION: Mainly upland sites that are moist in spring, in sagebrush steppe, short-grass prairies and open forest; tolerant of alkaline soils; E of the Cascades in OR and WA. BC to SK, south to CA, NM, and NE. Introduced to MO.

IDENTIFICATION TIPS: The rhizomatous habit and broad, shaggy, pale, unisexual inflorescences with tangled stigmas make *C. douglasii* easy to identify. It sometimes grows near *C. praegracilis,* which has blackish plant bases, thicker rhizomes, and narrower ♀ inflorescences. *Carex duriuscula,* rare in the PNW, has shorter inflorescences, shorter perigynia, shorter styles, and anther awn tips without minute bristles.

COMMENTS: *Carex douglasii* is often a community dominant. In our area, it grows in pale, fine-textured soils that are moist in spring but become powdery when dry. It is often found in dirt roads in sagebrush or dry open forest habitats. This comparatively short plant thrives when taller, more palatable plants are removed by livestock grazing, and it is tolerant of compacted, alkaline soils. Therefore, it persists as one of the few common plants on overgrazed alkaline ranges. In the absence of more palatable vegetation, it can provide nutritious forage despite being short and tough.

top left: perigynia
top *right and center left:* pistillate inflorescences
center right: habit
bottom left: staminate inflorescence

Carex duriuscula C. A. Mey.

Common name: Spikerush Sedge
Section: Divisae   Key: B, H
Synonyms: *C. eleocharis* L. H. Bailey, *C. stenophylla*
    auct., non Wahlenb.

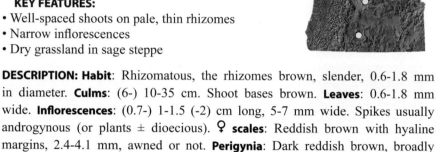

KEY FEATURES:
- Well-spaced shoots on pale, thin rhizomes
- Narrow inflorescences
- Dry grassland in sage steppe

DESCRIPTION: Habit: Rhizomatous, the rhizomes brown, slender, 0.6-1.8 mm in diameter. **Culms**: (6-) 10-35 cm. Shoot bases brown. **Leaves**: 0.6-1.8 mm wide. **Inflorescences**: (0.7-) 1-1.5 (-2) cm long, 5-7 mm wide. Spikes usually androgynous (or plants ± dioecious). ♀ **scales**: Reddish brown with hyaline margins, 2.4-4.1 mm, awned or not. **Perigynia**: Dark reddish brown, broadly elliptical to ovate or obovate, 2.4-3.9 mm long, 1.5-2.1 mm wide. Beak 0.3-0.9 mm long. Stigmas 2. **Achenes**: Lenticular. **Anthers:** Tipped with a tiny, relatively broad, smooth to warty awn.

HABITAT AND DISTRIBUTION: Dry prairie, sagebrush steppe, and open forest E of Cascades in OR (where last recorded in 1938). AK to YT, S to CA (White Mts.), NM and MO, also northern Asia and New Guinea.

IDENTIFICATION TIPS: *Carex duriuscula* is an inconspicuous, narrow-leaved sedge with thin, brown rhizomes and neat, oval inflorescences. *Carex douglasii* has similar rhizomes but is dioecious and has longer, wider inflorescences, larger perigynia, and anthers tipped with minutely bristly awns.

COMMENTS: *Carex duriuscula* can be a community dominant in dry grassland in parts of the Great Plains and northward. It has shifted its range N with the retreat of the glaciers, leaving small relict populations throughout the West. These populations are disappearing due to conversion to agricultural uses. Ironically, while declining in the west, *C. duriuscula* is expanding its range into eastern N America following roads that are salted in the winter.

Scientific names of plants are hypotheses about similarity and relationship. This sedge belongs to a confusing cluster of three taxa. If the slight differences between N American and Asian species were considered to warrant recognition at the species level, our plants would be called *C. eleocharis*. If the differences among N American, Asian, and European species were considered too trivial to distinguish species, all three would be united as *C. stenophylla*. We follow the current (but perhaps temporary) practice of treating N American plants as conspecific with Asian *C. duriuscula* but different from the European *C. stenophylla*.

Carex duriuscula

top left: perigynia
top right: inflorescences
bottom left: pistillate scale, achene, perigynia
bottom right: habit

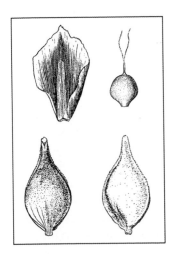

Carex eburnea Boott in W.J. Hooker
Common name: Bristleleaf Sedge
Section: Albae   Key: F

KEY FEATURES:
• Leaves thread-like
• Pistillate spikes raised on delicate stalks
• Staminate spike inconspicuous, whitish

DESCRIPTION: Habit: Cespitose but usually also producing rhizomes. **Culms**: 7-31 cm tall. **Leaves**: 0.2-1 mm wide, involute. **Inflorescences**: Lowest inflorescence bract virtually bladeless, with hyaline sheaths 2.5-8 mm long; lateral spikes ♀, 3-7 x 1.5-6 mm long, on long, delicate stalks mostly exceeding the inconspicuous ♂ terminal spike that consists of white-hyaline scales 2.5-4 mm long. ♀ **scales**: White-hyaline with green or brown midveins, 1-2 mm long, shorter than or equaling the perigynia, awnless. **Perigynia**: Light green maturing dark brown and glossy, 1.5-2.2 long x 0.7-1.1 mm wide, with beak 0.2-0.4(-0.5) mm long. Stigmas 3. **Achenes**: Trigonous. **Anthers** 1.3-1.8 mm long.

HABITAT AND DISTRIBUTION: Conifer and mixed forests, fens, sand dunes, rock outcrops, and tundra, on neutral or calcareous substrates; Pend Oreille Co., WA; AK to NB, S to WA, AR, and FL, disjunct in central Mexico.

IDENTIFICATION TIPS: The most difficult part of identifying *C. eburnea* is interpreting its unusual inflorescence. The pistillate spikes are easy to see, but the staminate spike, though technically terminal, is shorter than the stalks that hold the pistillate spikes and is easily misinterpreted as an aborted pistillate spike. With its delicate, involute leaves, *C. eburnea* resembles *C. anthoxanthea*, which has narrow, longer perigynia, and *C. disperma*, which has similar plump, short-beaked perigynia but sessile spikes and strongly rhizomatous growth habit.

COMMENTS: Ancestors of *C. eburnea* arrived in N. America over the Bering land bridge. *Carex eburnea* is closely related to *C. mckittrickensis*, a larger but otherwise similar plant that grows in west Texas. DNA analysis shows that it arose from *C. eburnea* ancestors. Therefore, recognizing *C. mckittrickensis* as a species leaves *C. eburnea* paraphyletic; it defines *C. eburnea* in a way that omits some of the members of the *C. eburnea* lineage, like defining a human family as including some but not all the grandchildren. Plant systematists prefer to avoid that, although paraphyly can be accepted at the species level. More disturbing, a few other, scattered *C. eburnea* populations are at least as distinctive as the west Texas plants in terms of morphology, genetics, or ecology. *Carex eburnea* forms a complex, a "metaspecies" that includes several more or less isolated populations that are evolving in their own ways. Expect botanists to disagree about how to classify them.

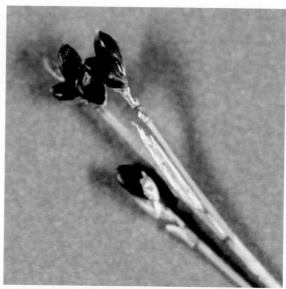

top left: perigynium
top right: inflorescence with 3 pistillate spikes and one slender, pale,
    staminate spike
bottom: habit

Carex echinata Murray ssp. *echinata*
Common name: Star Sedge
Section: *Stellulatae* Key: I
Synonyms: *C. muricata* L., misapplied; *C. angustior*
  Mackenzie, *C. ormantha* Mackenzie

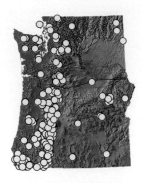

KEY FEATURES:
• Spikes short, "star shaped" with widely spreading
  perigynia
• Inflorescence elongate, spikes usually well separated

DESCRIPTION: Habit: Cespitose. **Culms**: 10-90 (-135) cm long. **Leaves**: usually 0.7-2.7 (-3.3) mm wide. **Inflorescences**: Spikes separated or sometimes crowded, distance between 2 lowest spikes usually longer than lowest spike. Spikes usually gynecandrous, with perigynia widely spreading or reflexed at maturity. Lateral spikes 3-15.5 mm long, terminal spikes 5-24 mm long, often with a narrow ♂ portion 2-16.5 mm long. ♀ **scales**: Inconspicuous, shorter than the mature achene. **Perigynia**: Green to brown, usually veinless on the ventral surface over the achene, with a swollen, pithy base when mature, (2.6-) 2.9-3.6 (-4) mm long, with beak 0.9-2 mm long, (35-) 38-60% of perigynium length. Perigynium tapered gradually to the beak. Stigmas 2. **Achenes**: Lenticular.

HABITAT AND DISTRIBUTION: Bogs, swamps, seeps, streamsides, wet meadows, usually in acidic soils but on serpentine in SW OR; widespread in the PNW but usually not coastal. AK to NF, south to CA and NC, but not in Great Plains; also Mexico, Central America, Dominican Republic, Hawaii, Australia, New Zealand, and Papua New Guinea.

IDENTIFICATION TIPS: *Carex echinata* ssp. *echinata* is a cespitose sedge with star-like spikes, often thriving with *Sphagnum* in montane bogs with little other herbaceous plant competition. Similar *C. interior* has short, stubby beaks only about 20-40 (-43)% as long as the perigynium bodies and tapering more abruptly, so the perigynia have "shoulders." In SW OR, *C. interior* has beaks at the longer end of that range, and the distinction can be difficult. Where the two grow together, *C. echinata* typically grows in seasonally wet, still water habitats; *C. interior* typically grows in sloping springs and seeps with moving water during all or part of the growing season. See *C. echinata* ssp. *phyllomanica*.

COMMENTS: *Carex echinata* ssp. *echinata* cover can be reduced following fire, perhaps because its growing points are often in a peat or moss layer that can burn. Plants may be older than they seem. European relative *C. muricata* may not flower until 3-4 years old, and may live for 20-35 years.

top left: perigynia   *top right and center left*: inflorescences   *bottom left*: habit

Carex echinata Murray ssp. *phyllomanica*
(W. Boott) Reznicek

Common name: Coastal Star Sedge
Section: *Stellulatae*   Key: I
Synonyms: *C. phyllomanica* W. Boott

KEY FEATURES:
- Spikes short, "star shaped" with widely spreading perigynia
- Inflorescence short, crowded
- Coastal bogs

DESCRIPTION: Habit: Cespitose. **Culms**: 20-80 cm long. **Leaves**: 1-3.3 (-3.8) mm wide. **Inflorescences**: Spikes crowded, distance between 2 lowest spikes shorter than lowest spike. Spikes usually gynecandrous, with perigynia widely spreading or reflexed at maturity. Lateral spikes 3-15.5 mm long, terminal spikes 5-24 mm long, often with a narrow ♂ portion 2-16.5 mm long. ♀ **scales**: Inconspicuous, shorter than the mature perigynia. **Perigynia**: Green to brown, usually with 2-12 veins on the ventral surface over the achene, with a swollen, pithy base when mature, (3.1-) 3.5-4.75 mm long, with beak 0.9-2 mm long, 33-45% of perigynium length. Perigynium tapers gradually to the beak. Stigmas 2. **Achenes**: Lenticular.

HABITAT AND DISTRIBUTION: Near the coast in bogs, deflation plains, swamps, seeps, streamsides, and lakeshores, AK to CA.

IDENTIFICATION TIPS: This is a densely cespitose sedge with widely spreading, long-beaked perigynia. The similar *C. echinata* ssp. *echinata* has narrower leaves, usually less crowded spikes, and shorter perigynia.

COMMENTS: In Alaska, *Carex echinata* ssp. *phyllomanica* has extremely long perigynia, crowded spikes, and wide leaves. Plants vary gradually toward the south. In the PNW, *C. echinata* ssp. *phyllomanica* more closely resembles ssp. *echinata*. Within the PNW there is considerable variation among *C. echinata* populations. Plants of NW CA and SW OR key to *C. echinata* ssp. *echinata* but have long perigynium beaks like ssp. *phyllomanica*, narrow leaves and separated spikes like ssp. *echinata*, and greater height than is typical of either variety. Around Mt. Hood, plants are typical of ssp. *echinata* except that they have the crowded spikes like ssp. *phyllomanica*. It is tempting to name these variations, but any taxonomic revision of *C. echinata* should result from a thorough study of the variation throughout its range, including centers of diversity in the Appalachians and the PNW.

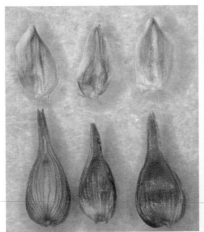

top left: pistillate scales,
   perigynia
top right: inflorescences
center left: inflorescences
   (green inflorescences
   with abnormal bracts
   are from late season
   shoots)
bottom left: habit

Carex engelmannii L. H. Bailey

Common name: Engelmann's Sedge
Section: *Inflatae*   Key: A
Synonyms: *C. breweri* Boott var. *paddoensis* (Suksdorf)
   Cronquist, *C. paddoensis* Suksdorf

KEY FEATURES:
• Single, fat spike
• Oval, flat perigynia much larger than the achenes
• High elevation, well-drained soils

DESCRIPTION: Habit: Rhizomatous, the rhizomes 1 mm thick. **Culms**: 7-15 cm tall. **Leaves**: Semi-cylindric or somewhat flattened, 0.3-0.6 mm wide, the leaf sheath fronts uniformly pale brown or colorless. **Inflorescences**: One thick, androgynous spike per culm, 0.7-1.5 cm long, 6-8 mm wide, lacking inflorescence bract. ♀ **scales**: Lower scales as wide and long as the perigynia; upper scales shorter and narrower, all 1-veined, yellowish brown, acute but involute near the tip and therefore appearing acuminate, not awned. **Perigynia**: Ascending, elliptic, veinless, 3.5-5 mm long, 1.5-2.5 mm wide, with rachilla longer than the achene and the achene much smaller than the perigynium. Stigmas 3. **Achenes**: Trigonous.

HABITAT AND DISTRIBUTION: Dry, open slopes above 6000 feet, often on pumice; in the Olympic Peninsula and the Cascades of WA S to Mt. Adams. S BC to MT, south to NV and CO.

IDENTIFICATION TIPS: *Carex engelmannii* is a short, rhizomatous, subalpine sedge with a single, large, dense spike and flat perigynia. *Carex breweri* differs in having ♀ scales with 3-5 veins and a whitish area around the midrib. *Carex breweri* also differs in having thicker rhizomes, leaves that are more circular in cross section, longer inflorescences, and longer, more nearly circular perigynia. *Carex breweri* and *C. engelmannii* are mainly allopatric but their ranges overlap on Mt. Adams, WA. *Carex subnigricans* is similar, but lives in moist habitats, and has a narrower inflorescence, wedge-shaped perigynium bases with a small stalk, and ♀ scales that are flat, not inrolled.

COMMENTS: Some authors have considered *C. engelmannii* and *C. breweri* to be a single species because the pistillate scale traits that distinguish the superficially similar species seem slight and size differences overlap slightly. However, differences in leaf anatomy and rhizomes are consistent with the inflorescence differences and suggest that the taxa really are distinct species.

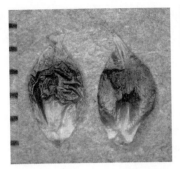

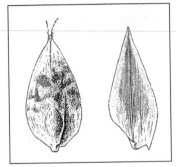

top left: perigynia
top right: inflorescence
center left: perigynium,
    pistillate scale
bottom right: habit

Carex exsiccata L. H. Bailey
Common name: Western Inflated Sedge
Section: *Vesicariae*   Key: E
Synonyms: *Carex vesicaria* L. var. *major* Boott in Hook.

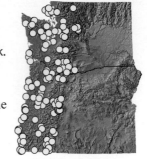

KEY FEATURES:
- Perigynia large and inflated, tapering gradually to the beak
- Leaf sheath nearly lacking crosswalls
- Cespitose

DESCRIPTION: Habit: Cespitose, or sometimes spreading on short rhizomes with crowded shoots. **Culms**: 30-100 cm tall. **Leaves**: Green, flat to W-shaped, the widest leaves 2.5-6.2 mm wide. Basal leaf sheath reddish, not spongy, nearly lacking crosswalls between the longitudinal veins. **Inflorescences**: Lateral spikes 2-5, ♀, erect or ascending, mostly 2-7 cm long, well separated from the 2-3 ♂ terminal spikes, 2-7 cm long. ♀ **scales**: Shorter than the perigynia, apex acute to acuminate, awnless. **Perigynia**: Ascending, green to straw-colored or reddish brown, with 9-20 veins, somewhat leathery, 7.5-10.1 mm long, 1.5-2.4 (-2.7) mm wide, tapering gradually to the beak 1.5-3 mm long, with straight teeth 0.2-0.9 mm long. Stigmas 3, styles persistent, becoming curved. **Achenes**: Trigonous.

HABITAT AND DISTRIBUTION: Pond margins, lake shores, wet meadows, and seasonal ponds where water pools in winter, generally in shallow water, 0-5400 feet elevation; common W of the crest of the Cascades in OR and WA, but occasional on the E side. BC to MT and CA.

IDENTIFICATION TIPS: *Carex exsiccata* is a large, cespitose sedge with large perigynia that taper gradually to a poorly defined beak with short, straight teeth. Very similar *C. vesicaria* grows E of the Cascades and has shorter perigynia that contract relatively abruptly to the beak. Immature plants are distinguishable only by geography. Many plants along the crest of the Cascades are intermediate. Similar *Carex utriculata* is strongly rhizomatous and has perigynia that taper more abruptly to the beak. It is more easily identified by the sides and backs of its leaf sheaths, which have a "brickwork" of crosswalls between the longitudinal veins. Where both species are present, *C. utriculata* grows in deeper water than *C. exsiccata*.

COMMENTS: *Carex exsiccata* can be a community dominant, well suited for wetland restoration W of the Cascades. Before the dams were built, plants along the Columbia survived 3-4 months underwater. *Carex exsiccata* is palatable to livestock, and in some areas is cut for hay. Rhizomes have been used for basketry. Leaves were used in Portland, OR, to make coarse ropes for handling newly cast pipes. Roots can be a source for a black dye.

top left: perigynia
top right: inflorescence
center left: spike
center right: habitat
  (typical tufted form)
bottom left: habitat
  (dark green swath,
  rhizomatous form)

Carex feta L. H. Bailey
Common name: Green-sheath Sedge
Section: *Ovales*   Key: J

KEY FEATURES:
- Cespitose, with gynecandrous spikes and winged perigynia
- Leaf sheath front green
- Erect, somewhat elongated, tidy plants
- Low to moderate elevations W of Cascades

DESCRIPTION: Habit: Cespitose. **Culms**: 50-100 cm tall. **Leaves**: 15-25 (-45) cm long, (2.5-) 3-4 (-5) mm wide. Leaf sheath front green and veined throughout, except for a white-hyaline triangle to 6 mm long near the top. Leaf sheath front may be prolonged 1.5-6 mm above the junction of the leaf blade with the sheath. **Inflorescences**: 3.5-8 cm long, 10-13 mm wide, elongated but with overlapping spikes, green to straw-colored. Spikes ovate, gynecandrous. ♀ **scales**: White hyaline or straw-colored, with midrib green, straw-colored, or brown, 2.7-3.5 mm long, shorter and narrower than the perigynia. **Perigynia**: Green, white, or straw-colored, the body ovate to broadly ovate, 3.2-4.2 mm long, 1.7-2.1 mm wide, with 1-8 veins on each face, wing 0.3-0.5 (-0.6) mm wide. Beak white or straw-colored at tip, winged and ciliate-serrulate; (1.5-) 1.8-2.4 mm from top of achene to tip of beak. Stigmas 2. **Achenes**: Lenticular, (1.3-) 1.5-1.8 mm long, 0.9-1.2 mm wide, 0.4-0.5 mm thick.

HABITAT AND DISTRIBUTION: Wet meadows and prairie, margins of marshes, and road ditches in and W of the Cascades. BC to CA, inadvertently introduced to Grant Co., OR, and Grand Teton National Park, WY.

IDENTIFICATION TIPS: The neat, tidy appearance, overall green inflorescence, green leaf sheath fronts, and low elevation W-side range identify *Carex feta*. See the similar *Carex fracta,* a coarser, sloppier plant with white-hyaline leaf sheath fronts and silvery inflorescences. *Carex scoparia's* leaf sheath front has such a narrow white-hyaline center that it can appear entirely green, but its usually curving inflorescence has a much finer texture with acuminate ♀ scales and pointed perigynia and its longer perigynia taper more gradually to the beak. See discussion of introduced *C. longii* and *C. tribuloides,* which have green sheath fronts.

COMMENTS: Wet prairie plant communities dominated by *C. feta* were probably common in the Willamette Valley at one time. *Carex feta* is used for marsh and wet prairie habitat restoration, mainly in the Willamette Valley. It was accidentally introduced to Grand Teton National Park during a wetland restoration project and is thriving there. Eradication efforts are underway.

top left: perigynia
top right: inflorescence
center left: leaf sheath top
center right: inflorescence
bottom left: habit (dense inflorescence is atypical)
bottom right: habitat

Carex filifolia Nutt. var. *filifolia*
Common name: Threadleaf Sedge
Section: *Filifoliae*   Key: A

KEY FEATURES:
- "Bunchgrass" with very narrow leaves, like Idaho Fescue
- Dense, brown, fibrous leaf bases
- Perigynia pubescent

DESCRIPTION: Habit: Densely cespitose, but plants sometimes growing together in dense stands that might appear rhizomatous. **Leaves**: Very narrow and wiry, 0.3-0.7 mm wide, with dense, persistent, brown sheaths. **Inflorescences**: One androgynous spike (1-) 1.2-2.6 (-3) cm long, with 5- 18 perigynia, lacking inflorescence bracts. ♀ **scales**: Reddish brown or yellowish with broad brownish or white margins, wider than the perigynia. **Perigynia**: Pubescent, at least on the upper part, obovoid, not compressed and sometimes definitely circular in cross section, (2.8-) 3-4.8 mm long, abruptly contracted to a beak 0.2-0.8 mm long, style base often conspicuously exserted from the beak. Rachilla well developed. Stigmas 3. **Achenes**: Trigonous.

HABITAT AND DISTRIBUTION: Sagebrush steppe on toeslopes, streamside terraces, and in dry meadows, with a little extra subsoil moisture so that the habitat is not fully xeric; usually below 7000 feet elevation (rarely to 8300 feet); E of the Cascades. AK to MN, south to CA and KS.

IDENTIFICATION TIPS: This distinctive sedge forms clumps that look very much like Idaho Fescue (*Festuca idahoensis*) or some other fine-leaved bunchgrass. The only similar sedge is the high elevation *C. nardina,* which has glabrous, flatter (planoconvex) perigynia with margins flattened and strongly serrulate distally. Carex elynoides, a rhizomatous Great Basin species that may occur in SE OR, has glabrous perigynia, does not have the style base conspicuously sticking out from the perigynium beak, and lacks a rachilla.

COMMENTS: *Carex filifolia* is a common, sometimes dominant, species of dry habitats. It provides excellent forage,with high protein levels in spring, for livestock, bighorn sheep, elk, and other species. Productivity declines with increasing grazing intensity. Plants survive fire well but recovery may be slow. Roots of well-grown plants spread to 30+ inches horizontally and to 60 inches deep. Seed production is low and establishment is probably episodic, occurring in wet years. Restoration projects are more successful when using plugs but improved methods are needed.

top left and middle: perigynia
top right: inflorescence
center left: habitat
bottom left: habit

Carex flava L.
Common name: Yellow Sedge
Section: *Ceratocystis*   Key: F

KEY FEATURES:
- Perigynia reflexed, with the beak bent back
- Inflorescence of 1 narrow ♂ spike above fat ♀ spikes
- Partial shade by lakes and reservoirs

DESCRIPTION: Habit: Cespitose. **Culms**: 10-75 cm tall. **Leaves**: 1.6-4.7 (-5.8) mm wide, green or yellow-green. **Inflorescences**: Lowest inflorescence bract much longer than spike, and often diverging at a right angle or more from the culm. Sheath lacking, or if present the lowest lateral spike is remote. Lateral spikes (1-) 2-5, ♀, 8-22 mm long, 7.5-12.7 mm wide, crowded or sometimes the lowest remote. Terminal spike ♂, sessile or pedunculate, 9-22 mm long, 1.1-3 mm wide. ♀ **scales**: Shorter than the perigynia. **Perigynia**: Reflexed, yellowish brown at maturity, strongly veined, 4-6.3 mm long, 1-1.9 mm wide, beak 1.3-2.7 mm long, bent back toward the dorsal side. Stigmas 3. **Achenes**: Trigonous.

HABITAT AND DISTRIBUTION: Mesic to damp edges of lakes and reservoirs, apparently establishing on eroding banks and other bare soil in the sun but persisting in partial shade of alders, salmonberries, and other woody plants; rare in N WA. AK to NF, S to WA, MT, IL, and NJ, also Europe and E to Iran.

IDENTIFICATION TIPS: *Carex flava* is a cespitose plant with a crowded inflorescence and reflexed perigynia with the beaks bent toward the dorsal side. It is similar to *C. viridula,* which has shorter, widely spreading but not reflexed perigynia with ± straight beaks 0.3-1.3 mm long. *Carex retorsa* may look similar in photos, but it is a much more robust plant with perigynia and spikes up to twice as long and its leaves twice as wide.

COMMENTS: *Carex flava* reaches the southern edge of its range in northern WA. It is an early successional species, thriving on the disturbance associated with fluctuating water. It does not resist trampling well. *Carex flava* is part of a complex of mostly European taxa that tend to hybridize when they grow together, but nonetheless remain more or less distinct. This species is sold for horticultural use, and a cultivar with yellow-streaked leaves has been developed.

As is true for many arctic and boreal sedges, development of flowering culms is a two-step process. The primordia of flowering culms are initiated during the short, cold days of fall. The flowering culms elongate and mature during long, warmer days the following spring. Unusually warm temperatures during culm initiation result in female flowers replacing males in part or all of the terminal spike.

top left: perigynia
top right:
    inflorescence
center left: habit
center right:
    inflorescences
bottom: habitat

Carex fracta Mack.

Common name: Fragile-sheath Sedge
Section: *Ovales*   Key: J
Synonym: *Carex amplectens* Mack.

KEY FEATURES:
- Cespitose, with gynecandrous spikes and winged perigynia
- Leaf sheath front extends 3+ mm above junction of blade and sheath
- Large, lax, sloppy plant
- Erect, somewhat elongated, silvery inflorescence

DESCRIPTION: Habit: Cespitose. **Culms**: 60-115 cm tall. **Leaves**: 20-45 cm long, 4-7 mm wide. Leaf sheath white-hyaline for at least 1 cm from top, often throughout; white-hyaline front extends 3-9 mm above junction of leaf blade with sheath. **Inflorescences**: 3.5-8 cm long, 10-16 mm wide, elongated but with overlapping spikes, silvery to green or straw-colored. Spikes elliptical, gynecandrous. ♀ **scales**: White hyaline or straw-colored, with midrib green, straw-colored, or brown, (2.8-) 3.6-4.5 mm long, shorter or longer and narrower than the perigynia. **Perigynia**: Green to straw-colored, the body broadly lanceolate to ovate, 2.9-4 (-4.8) mm long, 1-1.7 mm wide, with 5-11 veins on each face, wing 0.1-0.2 (-0.3) mm wide. Beak white or straw-colored at tip, usually unwinged, brown, parallel-sided, and more or less entire for the distal 0.1-0.5 mm; 1.3-2 (-2.8) mm from top of achene to tip of beak. Stigmas 2. **Achenes**: Lenticular, 1.3-1.6 mm long, 0.8-1.2 mm wide, (0.3-) 0.4-0.5 mm thick.

HABITAT AND DISTRIBUTION: Seasonally damp places, open forest, and roadsides, often in partial shade, from foothills to around timberline; mainly in the Cascades and SW OR, but occasional in E OR. WA to CA.

IDENTIFICATION TIPS: *Carex fracta* is a big, coarse, montane sedge with tall, leaning culms, a "60 mph sedge" identifiable even as one drives past. Its inflorescence is whitish or silvery. Its overall look is somewhat sloppy, with many overlapping, elliptical spikes and loose leaf sheaths prolonged at the mouth. *Carex feta*, a neat, tidy green plant of lowlands, differs in having leaf sheaths entirely green and veined, except for a small white triangle to 6 mm long at the top of the front. In SW OR and NW CA, these species are sometimes difficult to distinguish. There, *C. fracta* may have partly green leaf fronts like those of *C. feta*, but nearly always with a longer white area near the top; when in doubt, go by the overall appearance of the inflorescences.

COMMENTS: *Carex fracta* is a large, dramatic sedge with a large root mass, potentially useful in the garden and for habitat restoration.

Carex fracta

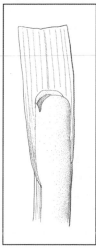

top left: perigynia
top right: inflorescence
center: top of leaf sheath front
bottom: habit

Carex geyeri Boott
Common name: Elk Sedge
Section: *Firmiculmes*   Key: A

KEY FEATURES:
- Single spike
- 1-3 large beakless perigynia
- Community dominant in dry, open, conifer forest

DESCRIPTION: Habit: Loosely cespitose or short rhizomatous, with thick, dark, shallow rhizomes. **Culms**: 12-50 cm, triangular in cross section. **Leaves**: Flat, 1.1-3.5 mm wide, evergreen, leathery, as long as or longer than the culms. **Inflorescences**: One androgynous spike per culm, lacking inflorescence bracts; ♂ terminal portion 1-2.5 cm long, separated from the 1-3 perigynia by a short internode. ♀ **scales**: ± brown with paler midrib and margins, pointed to awned, the lower ones longer than the perigynia, the upper ones reduced. **Perigynia**: Green, whitish, or brown, obovate, 5-7 (-8.5) mm long, 1.8-2.8 mm wide, rounded at the top, with a small beak. Rachilla present. Stigmas 3. **Achenes**: Trigonous.

HABITAT AND DISTRIBUTION: Well-drained soils in dry, open, conifer forest, aspen stands, and openings, rarely in sagebrush, on both serpentine and non-serpentine substrates, 1700 to 8200 feet; in mountains E of the Cascade crest, also SW OR. SE BC (rare) to AB, south to N CA and CO, disjunct and rare in PA.

IDENTIFICATION TIPS: *Carex geyeri* forms a ground cover in open forest. It produces tough, evergreen leaves from dark, scaly rhizomes that may be exposed at the soil surface. The inflorescences are inconspicuous, but the large perigynia are distinctive if present. If they have fallen, the pattern of 1-3 empty ♀ scales below a ♂ spike aids in identification. *Carex multicaulis,* found in SW OR, has a similar inflorescence except that its lower ♀ scales are green and leaf-like. It is densely cespitose and has round culms and involute leaves that are much shorter than the culms.

COMMENTS: Elk Sedge is one of the few *Carex* familiar to non-botanists. It is often a community dominant in dry, open, conifer forest, alone or with Pinegrass (*Calamagrostis rubescens*) or Idaho Fescue (*Festuca idahoensis*). An evergreen sedge, it provides valuable forage to diverse herbivores including wild and domestic ungulates, especially in winter. It holds soil well; the root mass may extend 1.4 m across and 1.8 m deep. The tough plants are resistant to trampling but vulnerable to soil disturbance by heavy equipment, which damages the shallow rhizomes. Usually, *C. geyeri* populations remain stable or increase after fire, regenerating from both rhizomes and buried seeds.

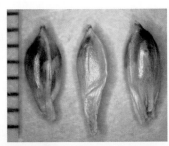
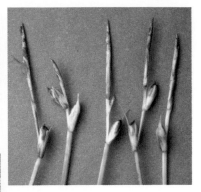

top left: perigynia
top right: inflorescences
center left: habit
bottom: habitat

Carex gynocrates Wormsk. ex Drejer

Common name: Northern Bog Sedge
Section: *Physoglochin*   Key: A, B
Synonyms: *C. dioica* L. ssp. *gynocrates* (Wormskjöld
ex Drejer) Hultén, *C. dioica* L. var. *gynocrates*
(Wormskjöld ex Drejer) Ostenfeld

KEY FEATURES:
• Single spike
• Perigynia spreading to reflexed
• Delicate plant of calcareous bogs

DESCRIPTION: Habit: Rhizomatous, the rhizomes only 0.3-0.8 mm in diameter. **Culms**: Slender and round in cross section, 2-30 cm long. **Leaves**: Very narrow, 0.3-0.7 mm wide. **Inflorescences**: One single spike that is usually entirely ♀ but may be androgynous or ♂. If ♂, the spike 0.8-1.6 cm long, 1-2 mm wide. If ♀, the spike more or less oblong, 0.5-1.4 cm long, 4-8 mm wide. ♀ **scales**: Light brown, more or less hyaline, mostly shorter but broader than the perigynia. **Perigynia**: Ascending at first but then spreading or even bending downward; yellowish or green when young, brown at maturity, with many obscure to conspicuous fine veins, oblong-ovoid, biconvex, 2.9-3.4 mm long, 1.2-1.7 mm wide, with a beak 0.5 mm long. Stigmas 2. **Achenes**: Lenticular.

HABITAT AND DISTRIBUTION: Growing in moss in bogs, usually in openings in conifer forest, sometimes in open bogs or subalpine meadows; moderate to high elevations; Okanogan Co., WA, and Wallowa Mts, NE OR. AK to NF, S to OR, UT, CO, and PA, also E Siberia.

IDENTIFICATION TIPS: *Carex gynocrates* is a delicate species winding through moss mats. Its solitary spikes and spreading perigynia can suggest *C. micropoda* or *C. nigricans,* but those species have consistently androgynous inflorescences and are much coarser. In addition, *C. micropoda* is cespitose and *C. nigricans* has leaves over 2 mm wide.

COMMENTS: The strong preference of *C. gynocrates* for calcareous substrates limits its potential habitat in the PNW, where it is at the southern edge of its range. Its wet habitat helps this species survive moderate fires but it is destroyed by intense fires that burn the moss mats in which it lives. This sedge is very similar to *C. dioica* of Europe and is often considered a subtaxon of that species. Most *Carex* with single spikes are most closely related to other single-spike species, but *C. gynocrates* and *C. dioica* are most closely related to species in subgenus *Vignea* which have multiple spikes.

top left: perigynia
top right: inflorescences
bottom left: habit
bottom right: habitat

Carex gynodynama Olney
Common name: Wonder Woman Sedge

Section: *Hymenochlaenae* (or *Porocystis*?)　　Key: C

KEY FEATURES:
- Hairy leaves and perigynia
- Floppy, cespitose plants

DESCRIPTION: Habit: Cespitose. **Culms**: 20-100+ cm tall. **Leaves**: 3-12 mm wide, flat to W-shaped, sparsely hairy, with ciliate margins. Lowest leaf sheaths reddish brown, bladeless, and hairy. Upper sheaths grade from dark red to green on back, with blades. Sheath fronts tan-hyaline, with or without red dots, usually pubescent at the top, sometimes hairy on the margins. **Inflorescences**: Brownish. Inflorescence bracts have well developed ± inflated sheaths 5-50 mm long. Lateral spikes 2-5, ♀, cylindrical, 1.2-4 cm long, 4-11 mm wide. Terminal spike ♂ (rarely gynecandrous), sessile or nearly so, 0.8-3 cm long. ♀ **scales**: Reddish brown with white margins and green midrib, broadly ovate, shorter than the perigynia, often hairy on midrib and awn. **Perigynia**: With rather long hairs, light green with reddish blotches at base and near tip, with up to 20 fine veins plus 2 marginal ribs, 3.7-5.3 mm long, 1-2.2 mm wide, with beak 1 mm long. Stigmas 3. **Achenes**: Trigonous.

HABITAT AND DISTRIBUTION: Seeps, moist meadows, open woodlands, and ditches, generally in disturbed sites in partial shade, often on serpentine-influenced soils (but not in severe serpentine substrates), at low elevations; Lane Co., OR, S to NW CA.

IDENTIFICATION TIPS: *Carex gynodynama* is a cespitose sedge with large, lax culms and hairy leaves and perigynia. The only other SW OR sedge with hairy leaves is *Carex whitneyi,* which has glabrous perigynia and lives at higher elevations. *Carex mendocinensis,* common within the range of *C. gynodynama,* has glabrous perigynia and glabrous or nearly glabrous leaves.

COMMENTS: This is a relatively ruderal species with a patchy distribution, sometimes forming large populations but usually observed as single plants or a few together, and often absent from apparently suitable habitats. *Carex gynodynama* and *C. mendocinensis* occasionally hybridize.

Carex gynodynama

top left: pistillate scales, perigynia

top middle and right : inflorescences (right with bug)

bottom left: habit

bottom right: inflorescence and leaf showing hairiness

Carex halliana L. H. Bailey
Common name: Hall's Sedge, Oregon Sedge
Section: Paludosae   Key: C
Synonyms: *Carex oregonensis* Olney ex L. H. Bailey

KEY FEATURES:
- Leaves leathery, stiff, arching
- Perigynia pubescent
- Well-drained, pumice soils in Cascades

DESCRIPTION: Habit: Rhizomatous, the shoots mostly arising singly. **Culms**: 10-40 (-50) cm tall, phyllopodic, with dull brown bases. **Leaves**: Green, tough, arching, 2-5.5 mm wide, with straw-colored or brown sheaths or the youngest with the sheath tinged reddish purple. **Inflorescences**: Bract of lowest lateral spike leaf-like and about as long as or longer than the inflorescence, with a somewhat inflated sheath (0.4-) 0.6-1.5 cm long. Lateral 2-4 spikes ♀ (rarely androgynous), erect to ascending, 1-5 cm long, much longer than wide. Terminal 1 (-3) spikes ♂, 1-3.5 cm long. All spikes crowded or the lowest remote. ♀ **scales**: Typically shorter than the perigynia, green with hyaline or brownish margins, glabrous or with margins scabrous apically. **Perigynia**: Ascending, ovoid, pubescent, 3.5-5.5 mm long, 1.6-2.2 mm wide, abruptly contracted to the beak. Beak 1.2-1.7 mm long, with straight teeth 0.2-0.6 mm long. Stigmas 3. **Achenes**: Trigonous.

HABITAT AND DISTRIBUTION: Well-drained soils, usually of volcanic origin, in meadows, open forests, and roadsides in the Cascades from Mt. Adams, WA, through OR. Disjunct and rare in CA from NE of Mt. Shasta to Medicine Lake, Siskiyou Co.

IDENTIFICATION TIPS: Compared to most other upland sedges, *C. halliana* is a husky, coarse plant, with shoots arising singly from rhizomes and spikes much longer than wide. Other fuzzy-fruited upland sedges such as *C. inops, C. brainerdii,* and *C. rossii* have lateral spikes slightly longer than wide. *Carex halliana* grows with *C. inops* ssp. *inops,* a much more delicate plant with a single ♂ spike, shorter ♀ spikes, and softer leaves.

COMMENTS: *Carex halliana* has been used to revegetate roadsides and trail margins in high elevation open forest and clearings in the Cascades, using plugs grown from seed. On its own, it colonizes cinders deposited on roadsides from gravel spread on snowy roads.

top left: perigynia
top right:
inflorescence
center left: spike
bottom: habit,
habitat

Carex harfordii Mack.

Common name: Monterey Sedge
Section: *Ovales* Key: J
Synonyms: *Carex montereyensis* Mackenzie

KEY FEATURES:

- Cespitose, with gynecandrous spikes and winged perigynia
- Lowest inflorescence bract often elongated
- Perigynia thick
- Coastal wetlands

DESCRIPTION: Habit: Cespitose. **Culms**: (30-) 50-120 cm, sometimes falling over, often branched late in the season. **Leaves**: Usually 2-5(9.9) mm wide, leaf sheath extending 1-4 mm above the base of the blade. **Inflorescences**: 1.5-4 cm long, 7-20 mm wide, dense and head-like, lowest internode <1.5-2.5 mm; lower inflorescence bracts usually short and bristle-like but sometimes leaf-like and longer than the inflorescence. Spikes gynecandrous. ♀ **scales**: Green to dark brown, center often green to gold, (2.5)3.5-4.7 mm long, usually shorter and narrower than the perigynia but sometimes longer and covering them. **Perigynia**: Green to brown, with metallic sheen, ascending to spreading-ascending, margin usually green, ovate to broadly ovate, (2.6-) 3.3-4 (-4.6) mm long, (1.3-) 1.4-1.6 (-2) mm wide, 0.6-0.7 mm thick, with walls +/- thin; with 4-9 veins on the dorsal surface and 0-9 veins on the ventral surface; veins usually exceeding the achene; wing 0.2-0.3 (-0.4) mm wide. Beak tip brown, cylindric, unwinged, and more or less entire for more than 0.4 mm from tip, 1.2-1.8 mm from achene top to beak tip. Stigmas 2. **Achenes**: Lenticular, (1.3-) 1.5-2.1 mm long, 0.9-1.4 mm wide, 0.5-0.7 mm thick.

HABITAT AND DISTRIBUTION: Wet or seasonally wet shores, meadows, and open forest near the coast, less than 2700 feet elevation, SW OR; CA.

IDENTIFICATION TIPS: *Carex harfordii* resembles *C. subbracteata* and unusually dark *C. pachystachya* with less spreading perigynia, but differs from both in having ventral perigynium veins that exceed the achene, like those of montane *C. abrupta*. Individuals with leaf-like inflorescence bracts might be confused with paler *C. athrostachya,* which has narrower, flatter perigynia.

COMMENTS: Putative *C. harfordii/subbracteata* hybrids from S CA are sterile. In *Carex* section *Ovales,* frequent self-pollination allows individuals with rearranged, holocentric chromosomes to reproduce, establishing whole lineages of plants with similar chromosomes and, in most cases, very similar morphology. This can result in a proliferation of very similar species, as in Carex Section *Ovales*.

Carex harfordii

top left: perigynia
center left: whole plant
other photos: variation in inflorescences

Carex hassei L. H. Bailey

Common name: False Golden Sedge
Section: *Bicolores*   Key: F, G, I
Synonyms: *Carex garberi* Fern (as used in the PNW)

KEY FEATURES:
- Perigynia pale
- Perigynium beak short and turned to the back
- Plants rhizomatous with pale leaf bases

DESCRIPTION: Habit: Rhizomatous, the rhizomes about 1 mm in diameter. **Culms**: 10-85 cm, averaging 35 cm tall. **Leaves**: Green to glaucous, 1.5-25 cm long, 2-3 (-3.7) mm wide. Plant bases whitish or pale to medium brown. **Inflorescences**: Lateral spikes ♀, (4-) 10-23 mm long, 3-5 mm wide, with perigynia densely or loosely spaced within the spikes, with middle internodes 0.2-1.1 mm, averaging 0.5 mm. Terminal spike ♂ or less often gynecandrous, when ♂ 0.6-2.8 cm long, averaging 1.3 cm long, (1.8) 2-2.5 mm wide. ♀ **scales**: 2-3.2 mm, brown, rounded or obtuse to acute, sometimes awned, usually persistent. **Perigynia**: Usually obovate or sometimes elliptical, with (4-) 9-15 veins, 1.8-3.2 mm long, 1.1-1.9 mm wide, pale green, whitish, or tan. Perigynia beakless or nearly so, the beak or perigynium tip usually curved to the back. Stigmas 2 (or a mix of 2 and 3 in the same inflorescence). **Achenes**: Lenticular (or trigonous).

HABITAT AND DISTRIBUTION: Uncommon in springs, fens, drier margins of marshes, and mesic or seasonally moist meadows, on various substrates including serpentine, from the coast to fairly high elevations; in the PNW mostly W of the Cascade crest. BC to CA, E to NV and perhaps CO.

IDENTIFICATION TIPS: *Carex hassei* is an inconspicuous, grass-like sedge. Its pale greenish or tan perigynia have short tips that bend toward the back. It is very similar to *C. aurea,* which has ♀ scales that often fall before the perigynia, and entirely succulent, orange (or white-powdery) mature perigynia that dry dark brown. Immature plants cannot be reliably distinguished, but in *C. aurea* usually the terminal ♂ spike averages smaller. In SW OR, *C. klamathensis* is similar but consistently has 3 stigmas per flower (except on aborted perigynia).

COMMENTS: Taxonomy of *Carex hassei* and relatives is controversial. It has often been merged with *C. aurea* and/or northern and eastern *C. garberi.* Even treated as a separate species*, C. hassei* has more variation than is typical of most *Carex* species. Variants include serpentine plants of NW CA and SW OR with a mix of 2 and 3 stigmas; small neat plants of alkaline springs in NV; and pale, robust plants of the San Bernardino Mts. of S CA.

top left: pistillate scales, dried perigynia with
    withered bases
top right: inflorescences
center right: fresh perigynia with orange,
    succulent bases
bottom left: habit
bottom right: habitat

Carex haydeniana Olney

Common name: Cloud Sedge
Section: *Ovales*   Key: J
Synonyms: *C. macloviana* d'Urv. ssp. *haydeniana*
(Olney) Taylor & MacBryde

KEY FEATURES:
- Cespitose, with gynecandrous spikes and winged perigynia
- Perigynia 4-6.5 mm long
- Compact, dark head with fine texture due to many, ± flat perigynia
- Subalpine to alpine habitat

DESCRIPTION: Habit: Cespitose. **Culms**: 9-40 cm. **Leaves**: 3-16 cm long, 1.5-4 mm wide. **Inflorescences**: 1.1-2.1 cm long, 13-18 mm wide, dense and head-like, with a fine texture because the perigynia are many and thin, not spreading. Spikes gynecandrous. ♀ **scales**: Generally brown or purplish, sometimes with a paler midstripe, 3-4.8 mm long, shorter and narrower than the perigynia. **Perigynia**: Light to dark brown (occasionally greenish) with dark to black tips, flat except over the achene, with several light dorsal veins and 0-3 (-8) ventral veins, 4-6.5 mm long, 1.5-2.6 mm wide. Wings 0.3-0.6 (-0.8) mm wide. Beak tip unwinged, brown, and parallel-sided for at least the distal 1 mm, entire for the distal 0.3-0.6 mm; (2.3-) 2.6-3.8 mm from achene top to beak tip. Stigmas 2. **Achenes**: Lenticular, (1.2-) 1.4-1.8 mm long, 0.8-1.1 (-1.3) mm wide, 0.3-0.5 mm thick.

HABITAT AND DISTRIBUTION: Moist or mesic, gravelly or rocky, subalpine or alpine slopes and flats, often in snowmelt zones; on Steens Mt., OR, and the higher peaks of NE OR. BC to AB, south to CA and CO.

IDENTIFICATION TIPS: *Carex haydeniana* is a short, cespitose, high elevation sedge with dense, black or blackish-green heads. Similar *Carex microptera* grows at moderate elevations to near timberline and is usually taller with shorter perigynia (3.4-5.2 mm long) that usually have green bodies, though they mature brown. Some individuals are intermediate. *Carex haydeniana* on Steens Mt. has perigynia with strong veins like those of *C. abrupta* but the shape of *C. haydeniana*.

COMMENTS: *Carex haydeniana* grows fast, a necessity for establishing itself in a cold, drying climate with a brief growing season. It can bloom a month after snowmelt during its second year. It is used for habitat restoration and erosion control at high elevations. Establishment is better from plugs than from seeds, especially because germination is reduced by the shade of erosion control blankets.

Carex haydeniana

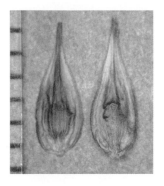

top left: perigynia
top right: inflorescences
center left: habitat
bottom left: habit

195

Carex hendersonii L. H. Bailey
Common name: Timber Sedge
Section: *Laxiflorae*   Key: F

KEY FEATURES:
- Cespitose with broad, soft leaves
- Inflated inflorescence bract sheath
- Moist forest and woodland

DESCRIPTION: Habit: Cespitose. **Culms**: 45-90 cm long, sprawling, floppy. **Leaves**: Dark green, 3-16 mm wide, with inflated sheaths. **Inflorescences**: Bracts with inflated sheaths. Spikes 1.2-4 cm long, widely separated. ♀ **scales**: Whitish with green midrib, 2.7-3.1 mm long, acute to awned. **Perigynia**: Green, trigonous, with many veins, 4.6-6 mm long, with the vaguely-defined inflated beak bending to the dorsal side. Stigmas 3. **Achenes**: Trigonous.

HABITAT AND DISTRIBUTION: Shaded riparian zones and moist forest; W of the Cascade crest in WA and OR, also SE WA. SW BC to CA, disjunct in SE WA and ID.

IDENTIFICATION TIPS: *Carex hendersonii* is a cespitose plant of moist forest, with broad leaves, inflated leaf sheaths, sprawling flowering culms, and widely separated spikes. In most moist forests W of the Cascades, only two sedges are common; *C. hendersonii* and *C. leptopoda*. *Carex leptopoda* and its relatives differ markedly by their narrower, paler green leaves, sheathless inflorescence bracts, shorter spikes, and lenticular perigynia.

COMMENTS: *Carex hendersonii* is dependent on adequate moisture and shade. After logging, surviving plants may form robust clumps but soon populations decline due to death of established plants and lack of recruitment. After low-intensity fires, *C. hendersonii* populations may recover to pre-fire levels within 2 years. Seedlings establish well in the burned forest floor. High-intensity fires have severe and long-lasting effects on populations because plants root mainly in the organic soil layer and duff. Disjunct populations in ID are considered sensitive due to their small size and narrow habitat requirements. *Carex hendersonii* seedlings are sold for use in moist, partly shaded situations in gardens. *Carex hendersonii* was named after Louis Henderson, a Professor of Botany at the University of Idaho from 1893 to 1908 and Curator of the University of Oregon Herbarium from 1924 to 1939. He was important for his numerous, well-documented collections and for encouraging other productive botanists.

top left: pistillate scales, perigynia
top middle: spike
top right: inflorescence
bottom left: habit
bottom right: staminate terminal spike above a pistillate
    spike

Carex heteroneura W. Boott
Common name: Different-nerved Sedge
Section: *Racemosae*   Key: F
Synonyms: *Carex atrata* L. var. *erecta* W. Boott,
   Carex epapillosa Mackenzie

KEY FEATURES:
• Spikes erect, lower ones peduncled, entire
   inflorescence nodding with age
• Spikes with contrast between dark scales and light
   perigynium margins
• Mesic to moist montane to alpine habitats

DESCRIPTION: Habit: Cespitose. **Culms**: 25-60 (-100) cm tall. **Leaves**: 2-6 mm wide. **Inflorescences**: 3-7 (-8) spikes, each 1-2.7 cm long, the lateral spikes ♀ and spreading to erect, the lowest spike sometimes separated from the others and pendent, the terminal spike gynecandrous. ♀ **scales**: Dark purplish black, shorter than to longer than the perigynium, tip acute or somewhat acuminate, the midrib pale or dark like the rest of the scale, margins hyaline, mottled, or pale. **Perigynia**: Elliptical to obovate or orbicular, more or less flattened except over the achene, green, yellow, or light to dark brown, 2.5-4.5 mm long, 1.75-3 mm wide, with a small beak 0.3-0.5 mm long. Stigmas 3. **Achenes**: Trigonous, filling half or less of the perigynium body.

HABITAT AND DISTRIBUTION: High montane to alpine streamsides, seeps, and moist to mesic meadows, and open forests, sometimes in wind-swept sites that have little snow accumulation, but not in fully dry sites; with very discontinuous distribution in the mts., in WA Cascades, and in OR in the Cascades and mts. of E OR. BC to SK, south to CA and CO.

IDENTIFICATION TIPS: *Carex heteroneura* is a cespitose, alpine sedge with spikes erect (except occasionally the lowest). The entire inflorescence often nods as the perigynia mature. Perigynia are broad and flat. Northern plants have a close superficial resemblance to *C. atrosquama,* which has elliptical, golden perigynia that are papillose distally, and obtuse ♀ scales. *Carex spectabilis* is similar but its spikes lack the contrast between dark ♀ scales and light perigynium margins, its spikes are less crowded, and it is a finer textured plant.

COMMENTS: *Carex heteroneura* can be a community dominant in forb-rich, high-elevation meadows. It has been recommended for high-elevation habitat restoration. Local variation in plant morphology suggests a lack of gene flow between populations in different mountain ranges. Some authors have treated several North American alpine taxa including *C. heteroneura* as parts of circumboreal *C. atrata. Carex heteroneura* is often treated as having two varieties, but they are not separable in OR and WA.

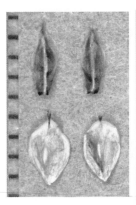

top left: pistillate scales and perigynia ("var. *heteroneura*")
top middle and right: inflorescences
center left: pistillate scales and perigynia ("var. *epapillosa*")
bottom left: habit
bottom right: habitat

Carex hirsutella Mackenzie
Common name: Hairy-leaved Sedge, Fuzzy Wuzzy
   Sedge
Section: Porocystis   Key: F
Synonyms: *Carex complantata* Torr. & Hook. var. *hirsuta*
   (L.H. Bailey) Gleason

KEY FEATURES:
- Perigynia beakless
- Leaves and sheaths hairy
- Terminal spike gynecandrous

DESCRIPTION: Habit: Cespitose. **Culms**: 20-90 cm long, much longer than the leaves and sprawling when mature. **Leaves**: 1.5-4 mm wide, blades hairy on both sides with hairs usually 0.4-0.6 mm long and bent or wavy; sheaths hairy. **Inflorescences**: Lowest inflorescence bract sheathless or nearly so; lateral spikes ♀, 8-18 mm long x 3-5.5 mm wide, often crowded; terminal spike gynecandrous with at least half the flowers pistillate, 10-20(-25) mm long. ♀ **scales**: White or green to brownish, ovate, shorter than the perigynia, obtuse, sometimes with short, abrupt tips. **Perigynia**: Green, maturing whitish to brown or purplish, ascending to spreading, weakly several-veined; 2-3 mm long x 1.1-1.6 mm wide, glabrous; beak absent. Stigmas 3. **Achenes**: Trigonous, 1.6-2.1(-2.6) mm long. **Anthers**: 1.3-2.2 mm long.

HABITAT AND DISTRIBUTION: Wet prairie remants, seasonally moist meadows and open, seasonally wet woodlands; introduced to Linn Co. OR, at 500 feet elevation. Native in E N Am. from ON to ME, S to TX and GA.

IDENTIFICATION TIPS: This is a grass-like sedge with virtually beakless perigynia. The leaves and sheaths are hairy, but this distinctive trait is not obvious until one looks closely. Vegetative plants would be easy to pass off as *Carex* section *Ovales* or *Carex densa* unless the hairy leaves were noticed. *Carex pallescens* is similar but has ♂ terminal spikes. Similar *C. swanii* (Fern.) Mack., introduced to Hernando Island, SW B.C., has densely hairy perigynia and shorter (0.7-1.3 mm) anthers.

COMMENTS: This eastern species was discovered in OR in 2013. The population comprises hundreds of plants in a degraded wet prairire on the east side of the Willamette Valley. It is puzzling that *C. hirsutella* and its relatives *C. pallescens* and *C. swanii* have established populations in the PNW when so many other eastern *Carex* have not. *Carex hirsutella* is very similar to eastern *C. complanata*, which has glabrous to glabrescent foliage, and is sometimes treated as *C. complanata* var. *hirsuta*. Differences between the two taxa are small but apparently consistent and they have slightly different though overlapping habitats.

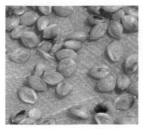

top left: perigynia
top right: inflorescences
center left: hairy leaves
bottom: habit (spawling fertile
    culms)

Carex hirta L.
Common name: Hairy Sedge
Section: *Carex*   Key: C

KEY FEATURES:
- Perigynia hairy
- Leaves and culms hairy
- Disturbed sandy soil near Portland

DESCRIPTION: Habit: Rhizomatous. **Culms**: (10-)20-90 cm tall. **Leaves**: 2.5-8 mm wide, hairy. **Inflorescences**: Lateral (1-)2-3 spikes ♀, erect or ascending; terminal 1-3 spikes ♂. ♂ **scales**: with short, scabrous awns (sometimes lacking on lowest scales), with sparse to dense, spreading, white hairs. ♀ **scales:** usually sparsely pubescent (occasionally glabrous), with scabrous awns. **Perigynium**: More or less densely pubescent with spreading hairs, lanceolate, with 12-20 veins, 4.8-7.8 mm long, 1.7-2.5 mm wide. Beak 1.5-2.7 mm long, with spreading teeth 0.8-1.7 mm long. Stigmas 3, style persistent. **Achenes**: Trigonous.

HABITAT AND DISTRIBUTION: Disturbed sandy soils, dry to wet meadows, stream banks, roadsides, and open forests; Portland, OR. Eurasia, introduced to E N America, New Zealand, OR.

IDENTIFICATION TIPS: This rhizomatous sedge with hairy perigynia resembles *C. sheldonii*, which is native E of the Cascades. *Carex sheldonii* differs in having glabrous ♀ scales and awnless ♂ scales that are glabrous or have sparse, +/- appressed hairs. *Carex pellita* has similarly hairy perigynia but glabrous leaves. Its perigynia are usually smaller and have shorter teeth.

COMMENTS: Completely glabrous forms of *C. hirta* exist in Eurasia, but have not been found in N. America. *Carex hirta* was found on ballast in Portland in 1916. It was not found again until 2010, when rediscovered among riprap on the shore of the Willamette River at the end of Ballast Street, appropriately enough. When cargo ships have little cargo, they take on ballast, heavy material used to lower a ship's center of gravity, thus preventing it from capsizing in high winds and improving its handling. At one time the ballast often consisted of sand or rocks. Inevitably, diverse plants and animals were transported with the ballast and deposited wherever a cargo was taken on. Portland's ballast dumps were combed by early botanists such as Wilhelm Suksdorf, and J. C. Nelson. These botanists made the first collections of many weeds that became established in the PNW, as well as unique records of species that failed to thrive here.

Carex hirta is the type species of the genus *Carex*, the species that anchors the definition of *Carex*. Therefore, no matter how much botanists may fiddle with sedge classification, *C. hirta* and its close relatives will always be in genus *Carex*.

top left: pistillate spike
top right: pistillate spikes, with terminal staminate
    spikes in background
bottom left: plant bases with rhizome
bottom right: hairy foliage

Carex hoodii Boott in W. J. Hook.
Common name: Hood's Sedge
Section: *Phaestoglochin*   Key: H, (J)

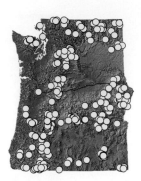

KEY FEATURES:
- Cespitose with androgynous spikes
- Perigynia green with coppery center
- Inflorescence a dense, rectangular to thimble-shaped head

DESCRIPTION: Habit: Cespitose. **Culms**: 20-80 cm tall. **Leaves**: Green, the widest 1-3.5 mm wide. **Inflorescences**: Dense heads 0.8-2 cm long, 6-15 mm wide, the spikes difficult to distinguish. Spikes androgynous. ♀ **scales**: Brown with a green midvein, 3.8-4.3 mm long, acute or short-awned. **Perigynia**: Shiny green with coppery or red-brown center, with age becoming light brown with dark brown center, veinless or with only obscure veins, 3.2-5 mm long, 1.4-2.5 mm wide, the margins serrulate distally, the body with marginal ribs but lacking flat winged margins, although the 0.7-1.5 mm long beak is flat-margined. Stigmas 2. **Achenes**: Lenticular.

HABITAT AND DISTRIBUTION: Mesic to dry montane grasslands, rocky hillsides, and openings in forest; widespread in the PNW, uncommon W of the Cascades, common in and E of the Cascades and in SW OR. BC to SK and NE, S to CA and CO.

IDENTIFICATION TIPS: Sooner or later the budding sedge-o-phile will attempt to key *C. hoodii* as an *Ovales*, and for good reason. However, this common, cespitose sedge is one of the most easily identified "*Ovales*-alike" sedges, distinguished by its shiny, veinless perigynia that have coppery centers contrasting with the green edges. (When old, they have dark brown centers and drab edges.) It is superficially similar to *C. microptera, C. pachystachya,* and other dense-head *Ovales,* but those species have winged perigynia (i.e., perigynia with flat margins) and gynecandrous spikes. *Carex jonesii* and *C. neurophora* perigynia have pithy bases, strong veins, and not much color contrast between perigynium center and edges.

COMMENTS: *Carex hoodii* can be an important forage for livestock and wildlife because it is common and moderately palatable, but it declines with heavy grazing. *Carex hoodii* also binds soil, preventing erosion. It is well suited for use in meadow restoration projects, roadside plantings, and soil stabilization projects.

Carex hoodii

top left: pistillate scales, perigynia
top right: inflorescences
center right: habit
bottom: habitat

Carex hystericina Muhl. ex Willd.
Common name: Porcupine Sedge
Section: *Vesicariae*   Key: E
Synonyms: *C. hystricina* Muhl. ex Willd.

KEY FEATURES:
• Nodding "bottlebrush" spikes with spreading, inflated
  perigynia
• ♀ scales with scabrous awns
• Cespitose

DESCRIPTION: Habit: Cespitose. **Culms**: 20-100 cm tall. **Leaves**: 2.5-8.5 mm wide, glabrous. Basal sheaths usually reddish purple. **Inflorescences**: Lateral 1-4 spikes ♀, 1.5-4 cm, erect or the lower ones nodding. Terminal spike ♂, 2-4 cm; ♀ **scales**: Lanceolate and tapering to a long, scabrous awn, shorter than the perigynia. **Perigynia**: Spreading or the lower ones reflexed when mature, yellow-green, inflated, glabrous, with 13-21 strong veins (the space between them usually 3+ times as wide as the veins), 4.5-7.3 mm long, 1.4-2.1 mm wide, with a beak 1.9-2.8 mm long with straight teeth 0.3-0.9 mm long. Stigmas 3, style persistent, becoming curved. **Achenes**: Trigonous.

HABITAT AND DISTRIBUTION: Streamsides in canyons and midslope seeps, usually in partial shade; E of the Cascades in WA and OR. Elsewhere, diverse wetlands, often without moving water, especially on calcareous substrates, S BC to NF, S to OR, NV (Jackson Mts.), TX, and VA.

IDENTIFICATION TIPS: *Carex hystericina* is an attractive, cespitose species with dangling spikes, and inflated perigynia with straight beak teeth. The scabrous awns on the ♀ scales distinguish it from most similar species except *C. comosa,* which has longer perigynium beak teeth that curve outward. *Carex vesicaria* and *C. utriculata* lack scabrous awns on the pistillate scales and have ascending pistillate spikes.

COMMENTS: In the PNW, *C. hystericina* persists where protected from grazing by steep slopes or narrow canyons. In the E and Midwest, *C. hystericina* is used in wetland habitat restoration projects, rainwater gardens, and infiltration basins. In the PNW, it has potential for restoration of ungrazed riparian zones. This showy plant can be an attractive addition to a moist garden. The species name was originally published as *C. hystericina*, but by 1886 the spelling was corrected to *C. hystricina*. If the name was intended to refer to the porcupine genus *Hystrix*, that correction is allowed under the International Code of Botanical Nomenclature, but the original author's intentions and thus the spelling remain controversial. Recently, some authors have reverted to *C. hystericina.*

top left: perigynia, pistillate scale
top right: inflorescence
center right: habitat
bottom right: habit

Carex idahoa L. H. Bailey

Common name: Idaho Sedge
Section: *Racemosae*   Key: A, F
Synonyms: *C. parryana* Dewey ssp. *idahoa* (L. H. Bailey)
  D. F. Murray

KEY FEATURES:
- Terminal spike long, with 0-3 small spikes below
- Plants rhizomatous, turf-forming
- Moist meadows in dry environments

DESCRIPTION: Habit: Rhizomatous, turf-forming but the stems sometimes arising in clusters. **Culms**: 25-40 cm tall. **Leaves**: 2-5 mm wide, slightly glaucous. **Inflorescences**: Terminal spike 2-3 cm long, 5-8 mm wide, solitary or with 1-3 short lateral spikes at its base. Plants often dioecious but sometimes with ♂ flowers scattered among the perigynia. ♀ **scales**: Brown to blackish with conspicuous paler midstripe and narrow pale margins, some scales 2-3 times longer than the perigynia, with a paler, conspicuous, often thickened midrib; tip acute to short-awned. **Perigynia**: Yellowish to brown, elliptic to obovate, 2-3 mm long, 1.5-1.75 mm wide, glabrous, veinless except for the 2 ribs, margins serrulate distally, with a small beak 0.2-0.3 mm long. Stigmas 3. **Achenes**: Trigonous, nearly filling the perigynium body.

HABITAT AND DISTRIBUTION: Flat to gently sloping meadows around headwater streams, ponds, and springs, often with *C. praegracilis* and/or *Poa pratensis,* in subirrigated soils that dry out on the surface in summer, in an ecotone between permanently wet areas and drier shrub steppe at mid to high elevations. Grant Co., OR, ID, MT, UT, and WY, disjunct in Mono Co., CA.

IDENTIFICATION TIPS: *Carex idahoa* is a glaucous, turf-forming plant with brown, usually solitary spikes. It may grow with *C. praegracilis,* which has more yellow-green leaves, darker plant bases and rhizomes, and a paler inflorescence with several short spikes. Closely related *C. parryana* differs in having lateral spikes about as long as the terminal spikes and ♀ scales about as long as the perigynia. It occurs near WA and OR in BC and ID. Its habitat is slightly drier and more alkaline.

COMMENTS: *Carex idahoa* is globally rare. Plants can be monoecious or dioecious, but ♂ plants are rarely collected. In ungrazed sites, it competes successfully with other graminoids. Grazing can cause increased competition from more grazing-tolerant, exotic pasture grasses.

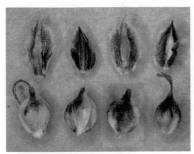

top left: pistillate scales, perigynia
top right: inflorescences
center right: habit
bottom left: habitat (mesic meadow)

Carex illota L. H. Bailey
Common name: Sheep Sedge
Section: undetermined   Key: I

KEY FEATURES:
- *C*espitose plant with gynecandrous spikes
- Dark, compact, pyramidal inflorescence
- Perigynium with a narrow rim rather than a wing

DESCRIPTION: Habit: Cespitose. **Culms**: 15-40 cm tall. **Leaves**: 4.5-18 cm long, 1.5-2.5 mm wide. **Inflorescences**: Pyramidal, 1-1.2 cm long, 5.5-10 mm wide, dense and head-like, blackish brown. Spikes gynecandrous. ♀ **scales**: Brown, usually with paler midrib, usually shorter and narrower than the perigynia. **Perigynia**: Gold to blackish brown, ovate, planoconvex, 2.6-3.2 mm long, 0.9-1.3 mm wide, unwinged but with 2 fairly sharp-edged, entire, lateral ribs, generally with some pithy tissue at the base. Beak unwinged, brown, and entire for the distal 0.4+ mm; 1.2-1.6 mm from achene top to beak tip. Stigmas 2. **Achenes**: Lenticular, 1.2-1.5 mm long, 0.8-1 mm wide.

HABITAT AND DISTRIBUTION: Montane marshes, streamsides, wet meadows, bogs; WA and OR. BC to SK, S to CA and CO.

IDENTIFICATION TIPS: *Carex illota* is a cespitose sedge with a dark, compact, often pyramidal inflorescence and wingless perigynia. It is similar to *C. integra,* which has very narrow perigynium wings and lives in drier habitats. The small, dark heads make *C. illota* superficially similar to small-headed individuals of *C. microptera,* but those plants have flatter perigynia with distinct wings. *Carex jonesii* perigynia bases are strongly swollen with pithy tissue.

COMMENTS: *Carex illota* may be a community dominant in some high elevation wet meadows in the Rocky Mountains, but in the PNW it is found as scattered plants. It is palatable to ungulates and where common is of some importance as forage. *Carex illota* has traditionally been considered a member of section *Ovales,* where it fits poorly because its perigynia are not winged. Recent research into evolutionary relationships among sedges indicates that it probably does not belong in section *Ovales*, and its sectional placement is uncertain.

top left: perigynia
 top right, center left and right: inflorescences
bottom: habitat

Carex infirminervia Naczi

Common name: Weak-veined Sedge
Section: *Deweyanae*   Key: I
Synonyms: *C. deweyana* Schwein. ssp. *leptopoda*
   (Mack.) Calder & Taylor, in part

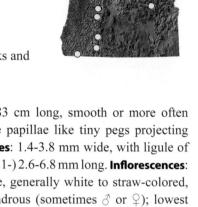

KEY FEATURES:
• Loosely cespitose, lax, grass-like leaves
• Moist forest
• Thin, membranous perigynia with long beaks and
   short teeth
• Ligules longer than wide

DESCRIPTION: Habit: Cespitose. **Culms**: 10-83 cm long, smooth or more often papillose (at 20X) at mid-length, the fragile papillae like tiny pegs projecting at right angles to the culm on its edges. **Leaves**: 1.4-3.8 mm wide, with ligule of uppermost culm leaf much longer than wide (2.1-) 2.6-6.8 mm long. **Inflorescences**: Spikes (4-) 5-6 (-7) usually longer than wide, generally white to straw-colored, smooth or rough in outline, usually gynecandrous (sometimes ♂ or ♀); lowest spike 9.5-18 mm long, with 11-22 perigynia; terminal spike 9.5-16 mm long. ♀ **scales**: 2.9-4.4 mm, silvery white to gold, with green midvein, usually awned, the awn to 1.1 mm long, the body about as long as or longer than the mature achene. **Perigynia**: 3.7-5.3 mm long, with beak 1.5-2.2 mm long, 39-49% of perigynium length, with teeth 0-0.2 (- 0.4) mm long. Stigmas 2. **Achenes**: Lenticular. **Anthers**: 1.3-1.8 mm long.

HABITAT AND DISTRIBUTION: Shaded to partially shaded riparian zones in mountains. Range uncertain because only recently recognized taxonomically, but occurring in the Cascades, NE WA, and N OR, probably widespread in mountains. BC to SK, S to CA and CO.

IDENTIFICATION TIPS: Several characters distinguish the "identical triplets," *C. bolanderi, C. infirminervia,* and *C. leptopoda. Carex infirminervia* is the one with long beaks but short beak teeth. Its culms have minute, peg-like papillae like *C. leptopoda.*

COMMENTS: At one time, botanists recognized two similar PNW species in this complex: *C. leptopoda* with short perigynium beaks and teeth, and *C. bolanderi* with long perigynium beaks and teeth. But what should plants with long beaks but short teeth be called? Confused botanists lumped all these plants together with northern *C. deweyana.* Recently, this group was tackled by a botanist who decided that (1) the PNW plants are definitely different from *C. deweyana,* which has short, few-flowered spikes and short ligules and (2) the long-beaked PNW plants with short teeth are a previously unrecognized species, now called *C. infirminervia.* Now PNW botanists are confused in a new way.

Carex infirminervia

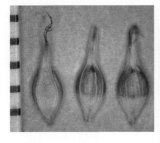

top left: perigynia
top right: inflorescences
center left: habit
bottom right: habit

Carex inops L. H. Bailey ssp. inops

Common name: Long-rhizome Sedge
Section: *Acrocystis*   Key: C
Synonyms: *C. pensylvanica* Lam. var. *vespertina* L. H.
   Bailey

KEY FEATURES:
• Pubescent perigynia
• Rhizomatous
• No basal spikes among leaf sheaths

DESCRIPTION: Habit: Rhizomatous. **Culms**: 13-50 cm tall, standing taller than the leaves. Plant bases brown or reddish. **Leaves**: Green, 0.7-4.5 mm wide, not papillose on the lower surface. **Inflorescences**: Bracts of lowest lateral spikes bristle-like or leaf-like, shorter than the inflorescence. Lateral spikes 1-4, ♀, each about 1.5 mm long or shorter, with 5-15 perigynia. Terminal spike ♂, 1-2.5 cm long, held above the ♀ spikes on peduncles (0.8-) 2.5-20 mm long. Basal spikes none (rarely with a basal ♀ spike on long, slender stalk). ♀ **scales**: Reddish brown, brown, or purplish with narrow white margins, about equaling the perigynia. **Perigynia**: Pubescent, obovoid, pale green to pale brown, with succulent bases that wither when dry, veinless except for 2 marginal ribs, 2.8-4.6 mm long, 1.5-2.2 mm wide; beaks 0.4-1.3 mm long with teeth 0.1-0.7 mm long. Stigmas 3. **Achenes**: Trigonous.

HABITAT AND DISTRIBUTION: Dry, open, conifer forest and meadows in the Cascades; also in well-drained, low elevation prairie and *Quercus garryana* savanna in the Puget Trough and San Juan Islands of WA, occasional in the Willamette Valley, OR. BC to Mt. Shasta, CA.

IDENTIFICATION TIPS: *Carex inops* is a rhizomatous upland species with pubescent perigynia in short ♀ spikes that are close to the ♂ terminal spike. It differs from *C. rossii* and *C. deflexa* var. *boottii* because those species are cespitose (sometimes loosely so), have basal ♀ spikes, and have leaf-like lowest inflorescence bracts longer than the ♀ spikes. *Carex halliana* is a stouter, coarser plant with stiff, leathery leaves, little-branched rhizomes, and lateral ♀ spikes much longer than wide.

COMMENTS: Because *C. inops* is common, can form a loose turf, and resists trampling, it has been recommended for revegetation in campgrounds and along trails, and as an element in native lawns. It has been recommended for use in "green roof" projects. *Carex inops* is best established from plugs, and should be mowed only occasionally. It is closely related to *C. pensylvanica* of eastern N America and is sometimes treated as a variety of that species.

top left: pistillate scales, perigynia
top right: inflorescence
center left: habit
bottom: habit, habitat

Carex integra Mack.
Common name: Smooth-beak Sedge
Section: *Ovales*   Key: I, J

KEY FEATURES:
- *C*espitose plant with gynecandrous spikes
- Fine-textured, light to brown, slightly elongate inflorescence.
- Perigynium with a very narrow wing
- Mesic to dry montane meadows

DESCRIPTION: Habit: Cespitose. **Culms**: (11-) 15-55 cm tall. **Leaves**: 5-18 cm long, (1.1-) 1.5-2.6 (-3.1) mm wide. **Inflorescences**: 1.1-2.4 cm long, 6.3-14 mm wide, dense and head-like, often slightly elongate. Spikes gynecandrous. ♀ **scales**: Gold to brown (to dark brown) with paler midrib, shorter than to slightly longer than the perigynia and about as wide. **Perigynia**: Gold to dark brown, lance-ovate to ovate or obovate, 2.1-3.6 mm long, 0.8-1.4 mm wide, smooth edged or nearly so, with a very narrow wing 0.05-0.2 mm wide. Beak tip unwinged, brown, and parallel-sided for at least the distal 0.4 mm; 1-1.6 mm from top of achene to tip of beak. Stigmas 2. **Achenes**: Lenticular, 1.1-1.4 mm long, 0.7-1 mm wide.

HABITAT AND DISTRIBUTION: Mesic to dry meadows at mid to high elevations in the mts; Cascades of OR and occasional E of the Cascades. OR to CA..

IDENTIFICATION TIPS: *Carex integra* is a cespitose sedge with a smallish head and smallish perigynia. It could be confused with *C. leporinella,* which has proportionately longer spikes and lives in soggy wet habitats. The perigynia resemble those of *C. illota,* which lives in more moist habitats, has a dark, short pyramidal inflorescence, and completely lacks perigynium wings; *C. integra* has a paler, longer inflorescence. It might be confused with *C. jonesii,* which has swollen perigynium bases, veined perigynia, and inconspicuously androgynous spikes, and lives in wetter habitats.

COMMENTS: *Carex integra* has softer foliage than most *Carex.* We have seen it eaten selectively by cattle. It is uncommon in most of its range, probably due to grazing, including the sheep grazing that was common in the western Cascades in the early 1900s.

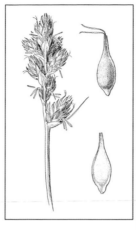

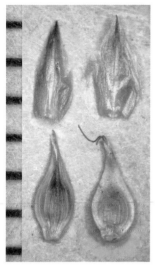

top left: pistillate scales, perigynia
top middle: inflorescence, perigynia
top right: inflorescence
bottom left: habit
bottom right: habitat (mesic to dry montane meadow)

Carex interior L. H. Bailey
Common name: Inland Sedge
Section: *Stellulatae*   Key: I

KEY FEATURES:
- Spikes separated, "star shaped," with widely
  spreading perigynia
- Terminal spike narrowed at base, club-like
- Perigynia contracting abruptly to short beaks

DESCRIPTION: Habit: Cespitose or short rhizomatous. **Culms**: 10-95 cm long.
Leaves: 1-2.4 (-2.7) mm wide. **Inflorescences**: Spikes separated, usually
gynecandrous (sometimes ♀), with perigynia widely spreading or reflexed at
maturity. Lateral spikes 3-9.5 mm long, terminal spikes 5.2-20 mm long, with
a narrow 2.2-14.5 mm long ♂ portion. ♀ **scales**: Inconspicuous, shorter than
the mature achene. **Perigynia**: Reddish brown to dark brown, sometimes green,
usually veinless on the ventral surface over the achene, with a somewhat swollen,
pithy base when mature, 1.9-3 (-3.3) mm long, beak margins usually serrulate,
with beak 0.4-0.9 mm long, 20-40 (-43)% of perigynium length. Perigynium
tapers relatively abruptly to the beak. Stigmas 2. **Achenes**: Lenticular.

HABITAT AND DISTRIBUTION: Springs and seepy canyon walls E of the Cascades
of WA and OR, and seeps (on serpentine or not) in SW OR. Perhaps introduced
to Tacoma, WA, where collected in 1901. AK to NF, S to CA, AZ, AR, and VA,
including the Great Plains.

IDENTIFICATION TIPS: *Carex interior* is an inconspicuous, slender-leaved, cespitose sedge with an interrupted inflorescence, the spikes consisting of widely
spreading perigynia that taper abruptly to the short beak, forming "shoulders."
Carex echinata has relatively longer beaks, 33-60% as long as the perigynia.
Perigynia of *C. echinata* taper gradually to the beaks. Where the two species
grow together, *C. interior* grows in shallow, usually moving, permanent water, for
example, a sloping seep, and *C. echinata* grows where water is usually still and
more seasonal. The habitat difference can be difficult to assess in late summer.

COMMENTS: Until recently *Carex interior* was thought to be rare in the PNW. It is
easily overlooked when growing with other grasses and sedges. It is now known
to be widespread but uncommon in its limited habitats.

top left: perigynia
top middle and right: inflorescences
center left: habit
bottom: habit

Carex interrupta Boeckeler

Common name: Greenfruit Sedge
Section: *Phacocystis*   Key: G

KEY FEATURES:
- Small, green perigynia without veins
- Lowest spike with widely spaced perigynia at its base
- Rhizomatous stands in sand along rivers

DESCRIPTION: Habit: Rhizomatous. **Culms**: 20-75 cm. **Leaves**: Narrow, 3-5 mm wide. Plant bases reddish brown or purplish black. Leaf sheath fronts hyaline, unspotted, not ladder-fibrillose. **Inflorescences**: Lowest inflorescence bract more or less equal to the inflorescence. Lateral 4-7 spikes ♀, green-and-blackish, long-tapering at base, with perigynia widely spaced at base of lowest spike, 4-9 cm long, 3-4 mm wide. Terminal 1 (-2) spikes ♂. ♀ **scales**: Blackish (or reddish brown), shorter than or equaling the perigynia. **Perigynia**: Green, not veined (occasionally with 1 or 2 irregular veins), 1.5-2.1 mm long, 0.9-1.4 mm wide. Beak 0.1-0.3 mm, not bidentate. Stigmas 2. **Achenes**: Lenticular.

HABITAT AND DISTRIBUTION: Seasonally flooded shorelines along rivers and major creeks, in sand and other coarse soils; sometimes along reservoir margins or in wet meadows, at low elevations (0-3000 feet). SW BC to SW OR, mainly W of the Cascade crest.

IDENTIFICATION TIPS: *Carex interrupta* is likely the correct name for any rhizomatous, narrow-leaved sedge of sandy soils along forested rivers W of the crest of the Cascades, particularly if its lateral spikes have widely spaced perigynia at their bases. This species may grow with *C. nudata,* which forms dense tussocks among riverine rocks or cobbles. On close examination, *C. nudata* has ladder-fibrillose leaf sheath fronts and larger perigynia. *Carex kelloggii* has small, green perigynia like *C. interrupta,* but *C. kelloggii* perigynia are veined, and the plants are usually cespitose.

COMMENTS: *Carex interrupta* was once considered a rare plant in the PNW, but recent field work revealed that it is more common than had been thought. However, it may be extirpated from BC. It is possible that dams have had a net negative effect on *C. interrupta* populations by reducing the flooding that creates its streamside habitat, although it occasionally colonizes the upper drawdown zones of reservoirs. It has potential to prevent soil erosion and should be considered for habitat restoration and soil erosion control projects in riparian zones in western OR and WA.

top left: pistillate scales, perigynia
top right: inflorescence
center left: habit
bottom left: habitat

Carex jonesii L. H. Bailey
Common name: Jones' Sedge
Section: uncertain; *Vulpinae*?   Key: H

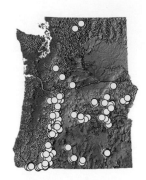

KEY FEATURES:
- Veined perigynia with swollen bases
- Dense head-like inflorescence
- Wet montane meadows, seeps

DESCRIPTION: Habit: Loosely cespitose. **Culms**: 15 to 40 (-60) cm tall, with basal sheaths of previous years persisting as fibers. **Leaves**: Green, 1-4 mm wide, all clustered near the base of the shoot, arising at about the same height so that the sheaths are mostly hidden. All or most sheaths with well developed blades. Leaf sheath fronts hyaline, usually smooth (occasionally somewhat cross-corrugated), shorter than the blades, with concave mouth. **Inflorescences**: Dense heads 0.8-2 (-2.5) cm long, 0.6-1.5 cm wide, with androgynous spikes (or some spikes ♀). ♀ **scales**: Dark brown, about as long as the perigynia. **Perigynia**: Widest near the base (occasionally near middle), brown, 2.5-3.5 (-4) mm long, 1-1.5 mm wide, strongly veined, with 7-11 veins on the dorsal surface, 5-7 on the ventral surface, the base is usually pithy and strongly swollen, the beak 0.5 to 1.5 (-2) mm long, with margins smooth (or nearly so). Stigmas 2. **Achenes**: Lenticular.

HABITAT AND DISTRIBUTION: Wet meadows, seeps, and streamsides in the mountains, 4000-8400 feet; in the Cascades and the mountains to the E in WA and OR, and in the Klamath Mts. in SW OR. WA to MT and CO, S to CA and NM but not UT or AZ.

IDENTIFICATION TIPS: *Carex jonesii* is a small, usually dark-headed sedge of montane wetlands, sometimes referred to as the Big Butt Sedge because of the appearance of its swollen, pithy perigynium bases. Its leaves are crowded near the base of the shoot, hiding the leaf sheath fronts. It is superficially similar to species of section *Ovales,* and *C. illota,* but those species lack swollen, pithy perigynium bases and they have winged perigynia (except *C. illota*) and gynecandrous spikes. *Carex illota* lacks the strong perigynium veins of *C. jonesii. Carex jonesii* often grows with the similar *C. neurophora* which has cross-corrugated leaf sheath fronts that are usually exposed because the leaf blades arise at different heights. In SW OR, see *Carex nervina.*

COMMENTS: *Carex jonesii* can be a community dominant in wet meadows. Although it is common in OR, there are relatively few WA records. The perigynia are an important food for White-tailed Ptarmigan in the Rockies. It is possible that *C. jonesii* is not closely related to the other members of *Carex* section *Vulpinae,* but rather to *C. illota* and/or *C. vernacula.*

top left: perigynia
top right: inflorescences
center left: habit
bottom: habitat

Carex kelloggii W. Boott var. *impressa*
B. L. Wilson & N. Otting

Common name: Mountain Shore Sedge
Section: *Phacocystis*   Key: G
Synonyms: *Carex lenticularis* Michx. var. *impressa* (L.H.
   Bailey) L.A. Standley

KEY FEATURES:
• Densely cespitose
• Green, stipitate perigynia with 1-3 veins
• Beaks and upper part of perigynia with brown markings

DESCRIPTION: Habit: Cespitose, or sometimes ± rhizomatous. **Culms**: 15-60 cm. **Leaves**: Green or slightly glaucous, 2-3.5 mm wide. Plant bases brown or reddish brown. Leaf sheath fronts hyaline, not ladder-fibrillose. **Inflorescences**: Lowest inflorescence bract usually longer than the inflorescence. Lateral 3-4 spikes ♀, 1.5-3.6 cm long, 3-4 mm wide. Terminal spike ♂. ♀ **scales**: Pale brown to black, shorter than the perigynia. **Perigynia**: Elliptic to ovate, stipitate, green with purplish brown spots on apical half, with 1-3 veins on the dorsal surface, 1.8-2.5 mm long, 1.1-1.5 mm wide. Beak 0.2-0.5 mm, reddish brown throughout or occasionally with some green, not bidentate. Stigmas 2. **Achenes**: Lenticular.

HABITAT AND DISTRIBUTION: Lakeshores, reservoir margins, pools along rivers, and other areas of quiet or slow moving water but often with fluctuating water levels, sometimes a community dominant above timberline; mainly in the Cascades, also the mts of E OR. WA to ID and CA.

IDENTIFICATION TIPS: This is a cespitose sedge with green, stipitate, lenticular perigynia, spikes > 1.5 cm long, and separate male and female spikes. Similar *C. kelloggii* var. *kelloggii* has perigynia that have 5-7 veins and are green except for a narrow bit of brown at the beak tip. *Carex aquatilis* is strongly rhizomatous, with wider leaves, and perigynia that are more purplish or brownish, veinless and rarely stipitate. *Carex nigra,* introduced to SW BC, is similar but is strongly rhizomatous.

COMMENTS: A *Carex kelloggii* clump may grow as a ring if shoots die out in the center. When many plants grow close together the cespitose habit may be obscured. *Carex kelloggii* var. *impressa* is valuable for erosion control along mountain lakes and streams. It commonly lines the upper margins of drawdown zones of reservoirs in the Cascades. In this setting, it survives total inundation for part of the year. Recent genetic work indicates that our western taxa should be separated at the species level from eastern *C. lenticularis*, which has paler plant bases.

Carex kelloggii var. impressa

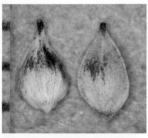

top left: perigynia
top right: inflorescences
center left: habit (cespitose *C. kelloggii* var. *impressa* in front; rhizomatous *C. obnupta* behind)
bottom: habit, habitat (reservoir drawdown zone)

Carex kelloggii W. Boott var. *kelloggii*

Common name: Kellogg's Sedge, Lakeshore Sedge
Section: *Phacocystis*   Key: G
Synonyms: *Carex lenticularis* Michx. var. *lipocarpa*
  (T. Holm) L.A. Standley

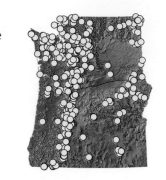

KEY FEATURES:
- Densely cespitose
- Perigynia with 5-7 veins, stipitate
- Perigynia green except for brown beak tip

DESCRIPTION: Habit: Cespitose. **Culms**: 15-90 cm. **Leaves**: Green or slightly glaucous, 2-4 mm wide. Plant bases brown or reddish brown. Leaf sheath fronts hyaline, not ladder-fibrillose. **Inflorescences**: Lowest inflorescence bract usually longer than the inflorescence. Lateral 3-7 spikes ♀, 1.5-3.6 cm long, 2.5-4 mm wide. Terminal spike ♂. ♀ **scales**: Pale brown to black, shorter than the perigynia. **Perigynia**: Elliptic to ovate, stipitate, whitish or green, with 5-7 veins on each face, 2-3.5 mm long, 1-1.8 mm wide. Beak 0.1-0.5 mm long, usually green with reddish brown at the very tip only, or rarely with a little brown extending down the beak, not bidentate. Stigmas 2. **Achenes**: Lenticular.

HABITAT AND DISTRIBUTION: Lakeshores, reservoir margins, pools along rivers, and other areas of slow water but often fluctuating water levels, usually in mountains throughout our region. AK and AB, S to CO, AZ, and CA.

IDENTIFICATION TIPS: This is a cespitose sedge with green, stipitate, lenticular perigynia and spikes > 1.5 cm long. It is the most common and widespread variety of *C. kelloggii* in the PNW. Similar *C. kelloggii* var. *impressa* perigynia differ in having a shorter stipe, brown coloration in the upper half of the body, and the beak tip brown throughout. *Carex kelloggii* var. *kelloggii* may grow with *C. aquatilis*, which is strongly rhizomatous and has wider leaves and veinless, non-stipitate, purplish or brownish perigynia.

COMMENTS: This species is often a community dominant. It can establish in a disturbed habitat and may remain indefinitely if the combination of saturated soil and its own dense growth exclude competing vegetation. Vertical rhizomes allow the plant to grow upwards through deposited sediments in disturbed habitats. *Carex kelloggii* var. *kelloggii* produces large numbers of seeds that germinate readily on soil exposed by falling water levels. Its extensive root system holds soil in an area wider and deeper than its foliage might suggest. It has been used for habitat restoration and erosion control in appropriate montane and alpine habitats. Because *C. kelloggii* remains green in fall and winter, this not particularly palatable species can be seasonally important forage for livestock and wildlife.

Carex kelloggii var. *kelloggii*

top left: pistillate scales, perigynia
top right: inflorescences
center left: habit
bottom left: habit, habitat

Carex kelloggii W. Boott var. limnophila (Holm) B. L. Wilson & R. E. Brainerd

Common name: Coastal Shore Sedge
Section: *Phacocystis*　　　Key: G
Synonyms: *C. hindsii* C. B. Clarke, *Carex lenticularis* Michx. var. limnophila (T. Holm) Cronquist

KEY FEATURES:
- Densely cespitose
- Green, stipitate perigynia with 5-7 veins
- Thick, crowded spikes

DESCRIPTION: Habit: Cespitose. **Culms**: (7-) 17-45 (-80) cm. **Leaves**: Green, 2-4 mm wide. Plant bases brown or reddish brown. Leaf sheath fronts hyaline, not ladder-fibrillose. **Inflorescences**: Lowest inflorescence bract usually longer than the inflorescence. Lateral 3-7 spikes ♀, 1.5-3.6 cm long, 4-6 mm wide, crowded and overlapping. Terminal spike ♂. **♀ scales**: Reddish brown to black, shorter than the perigynia. **Perigynia**: Elliptic to ovate, stipitate, green with purplish brown spots on distal half, with 5-7 veins on each surface, 2.5-3.5 mm long, 1.3-1.6 mm wide. Beak 0.2-0.5 mm, green with reddish brown rim at the very tip, occasionally with brown running down the beak, not bidentate. Stigmas 2. **Achenes**: Lenticular.

HABITAT AND DISTRIBUTION: Deflation plains, lake margins, pools, and other shallow wetlands at the coast, occasionally introduced to low elevation, inland sites W of the Cascades. AK to CA.

IDENTIFICATION TIPS: Coastal *C. kelloggii* var. *limnophila* has thick, overlapping spikes that are more than 1.5 cm long. It does not grow with the other *C. kelloggii* varieties, which have longer, narrower leaves and less crowded inflorescences. *Carex obnupta* is rhizomatous and taller, and has coarser, evergreen foliage.

COMMENTS: Although it can form large tussocks like its relatives, *C. kelloggii* var. *limnophila* more often grows as small clumps or single stalks. Its typical deflation plain habitat exists only briefly before shore pine and shrubs invade or another dune covers it. Dune stabilization due to the introduced European Beach Grass (*Ammophila arenaria*) is altering this habitat, placing this plant at risk. Like its relatives, *C. kelloggii* var. *limnophila* establishes in disturbed sites. Two inland, introduced populations are known, one persisting at an abandoned mill pond in Lane Co., OR, where *C. kelloggii* var. *limnophila* grows on floating logs.

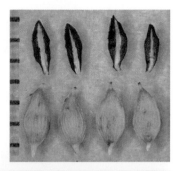

top left: pistillate scales, perigynia
top right and center left: inflorescences
bottom: habit, habitat

Carex klamathensis B. L. Wilson and L. P. Janeway
Common name: Klamath Sedge
Section: *Paniceae*   Key: F

KEY FEATURES:
- Rhizomatous sedge with glaucous foliage
- Stigmas 3
- *Darlingtonia* fens on serpentine substrates

DESCRIPTION: Habit: Rhizomatous, the rhizomes 1-2 mm in diameter. **Culms**: 30-100 cm tall, phyllopodic. **Leaves**: Glaucous, 18-50 cm long, flat, the wider leaves 2-6 mm wide. Plant bases whitish or pale to medium brown. **Inflorescences**: Lateral spikes usually 1-3, ♀, 1-2.5 cm long, 4-7 mm wide. Terminal spike usually ♂, but in some populations mostly gynecandrous, if ♂ 1.3-2.7 cm long, 1-5 mm wide, with 50-200 flowers. **♀ scales**: 1.9-2.8 mm long, gold to brown with paler midrib, obtuse, sometimes with an awn up to 1.5 mm long. **Perigynia**: Obovate or rarely fusiform, 2.1-3.6 mm long, 0.6-1.8 mm wide, light green, tan, or whitish, sometimes marked with brown distally; with 8-20 faint to strong veins, usually papillose at least near the short, bent beak. Stigmas 3. **Achenes**: Trigonous.

HABITAT AND DISTRIBUTION: Springs and fens on serpentine substrates, often with *Darlingtonia californica,* usually in a dense thatch of old leaves; Josephine Co., OR. Also at four locations in NW CA.

IDENTIFICATION TIPS: *Carex klamathensis* is an inconspicuous, glaucous, rhizomatous sedge of serpentine fens, with pale, veined, somewhat papillose perigynia that usually have the apex turned to the back. Glaucous *C. livida,* which lives only in non-serpentine bogs, has longer (3.7-4.8 mm) pale perigynia that are consistently spindle-shaped. Very similar *C. hassei* may grow in serpentine bogs but has 2 stigmas per flower or, in SW OR and NW CA, a mix of 2-stigma and 3-stigma flowers. (Occasionally *C. klamathensis* can seem to have a mix of 2- and 3-stigma flowers because aborted flowers have 2 stigmas.) The glaucous foliage of *C. klamathensis* resembles the rush *Juncus orthophyllus.*

COMMENTS: This globally rare sedge was discovered independently by botanists Peter Zika and Lawrence Janeway. They both initially reported it as *C. livida* but later realized that it was a distinct taxon. It is threatened by mining, road building, and recreational activities. Perhaps the few, widely separated populations are remnants of what was once a broad distribution in the Klamath Mountains.

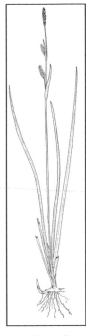

top left: pistillate scales, perigynia
top middle: habit
top right and center left: Inflorescences
bottom: habitat (serpentine fen)

Carex kobomugi Ohwi

Common name: Japanese Sedge
Section: *Macrocephalae*   Key: F, B

KEY FEATURES:
- Giant, spiky inflorescences
- Very large perigynia
- Disturbed sandy soil

DESCRIPTION: Habit: Rhizomatous. **Culms**: 10-30 cm tall, bluntly 3-angled, the culms smooth near the inflorescence. **Leaves**: 3-6 mm wide, longer than the culms but curving sideways or down, serrulate on margins. **Inflorescences**: Each culm with a single dense head consisting of many densely aggregated spikes. Each head either ♂ or ♀, the plant thus appearing dioecious, although ♂ and ♀ shoots are connected by rhizomes. ♂ inflorescences 3-4 cm long and 1-2 cm wide. ♀ inflorescences 3-6 cm long, 2-4 cm wide, with lower spikes ascending (to spreading), sometimes with inconspicuous ♂ flowers mixed in. ♀ **scales**: Long-tapered or sometimes with an awn 6-12 mm long. **Perigynia**: 10-14 mm long, green or yellow to dark brown, lance-ovate to elliptic, with wings that have irregular margins. Beak shorter than or about as long as the body, with distance from the top of the achene to the tip 3-5 mm. Beak tip notched 0.4-0.6 mm. Stigmas 3. **Achenes**: Rounded-trigonous. **Anthers:** 4-6.5 mm long.

HABITAT AND DISTRIBUTION: Sandy soils on the lower Willamette and Columbia rivers and nearby coastal sands; SW WA and NW OR. Introduced to the coast of E N American from MA to NC. Native to coastal Japan, Taiwan, and Korea.

IDENTIFICATION TIPS: This is a strongly rhizomatous sedge of sandy soils, with wide, tough, leaves with strongly serrulate margins, and large, dense heads with perigynia 1+ cm long. Similar *C. macrocephala* is native to coastal sand dunes. It has sharp angles on the culm, darker ♀ scales, anthers 2.5-5 mm long, and perigynia with a proportionately longer beak.

COMMENTS: *Carex kobomugi* was reported in Portland in 1916, in ballast sand from ships. It had appeared on the E coast by 1929, following a shipwreck that included Asian porcelain packed in its shoots and rhizomes. The "Sea Isle" cultivar has been planted in disturbed coastal sands of the E coast because its tough leaves resist trampling. Seed germination is low, but plants spread aggressively by rhizomes, excluding native species. Now E coast populations are being eradicated by digging the deep rhizomes and by herbicide application. Populations established in industrial districts of Portland appear to be the seed sources for an incipient invasion of coastal dunes in SW WA. Both achenes and rhizomes are nutritious for humans.

top left: perigynia
top right: inflorescence
left: habit
lower left: habitat

Carex lacustris Willdenow
Common name: Lake Sedge
Section: Paludosae   Key: E

KEY FEATURES:
- Strongly rhizomatous
- Perigynia inflated, short-beaked, brown
- Ligules long

DESCRIPTION: Habit: Strongly rhizomatous, forming monospecific stands. **Culms**: 50-136 cm tall, plant bases reddish, strongly ladder-fibrillose, lowest sheaths lacking blades. **Leaves**: Glaucous, (5.5-) 8.5-21 mm wide, glabrous, M-shaped in cross section, with longest ligules 13-40+ mm long, much longer than wide. **Inflorescences**: Lower 2-4 spikes ♀, ascending to arching, terminal 3-5(-7) spikes ♂. **♀ scales**: Glabrous, tip with awn 0.3-3.5 mm long, +/- scabrous. **Perigynia**: narrowly ovoid to ellipsoid, tough, strongly 14-28-veined, (4.5-) 5.2-7.8 mm long x 1.5-2.5 mm wide, glabrous, tapering gradually to a short beak 0.5-1.6 mm long, with 2 straight teeth 0.2-0.7(-0.9) mm long. Stigmas 3. **Achenes**: Trigonous.

HABITAT AND DISTRIBUTION: Swamps, marshes, wet sedge meadows, Pend Orielle Co., WA; AB to NF, S to ID and VA.

IDENTIFICATION TIPS: *Carex lacustris* forms extensive stands in wetlands. The stand may be sterile, and if it does have inflorescences they tend to be inconspicuous, mostly shorter than the vegetative shoots. It may grow adjacent to *C. utriculata*, which has greener leaves with ligules about as long as wide and perigynia that taper more abruptly to the beak. *Carex atherodes* has similarly long ligules, but its foliage is hairy at least at the leaf sheath mouth, and its perigynia are normally sparsely hairy.

COMMENTS: In 2013, fertile specimens of *C. lacustris* were collected in NE WA. Its presence there had been suspected for several years, but without fertile shoots it was impossible to identify the glaucous stands confidently. *Carex lacustris* hybridizes with *C. pellita* and *C. utriculata*. Hybrids are intermediate in morphology between the parents, and are sterile. Large stands of *C. lacustris* attract diverse herbivorous insects including leaf beetles, Sedge Billbugs, aphids, leafhoppers, spittlebugs, seed bugs, plant bugs, stem-boring larval flies, sedge grasshoppers, and caterpillars of skippers, brown butterflies, and moths. These in turn attract insectivorous animals such as dragonflies and birds. *Carex lacustris* seeds are eaten by waterfowl and other birds. Muskrats and deer eat the young foliage. Turtles eat the seeds or young foliage to a limited extent. The extensive stands provide excellent cover for many kinds of wildlife. *Carex lacustris* is planted for erosion control along margins of ponds and slow-moving streams.

top left: perigynia
top right: infloresence
center left: awned pistillate
    scales
bottom left: long ligule
bottom right: glaucous, rhizomatous stand (right) with yellow-green
    stand of *C. utriculata* (left)

Carex laeviculmis Meinsh.

Common name: Smoothstem Sedge
Section: uncertain; *Deweyanae?*   Key: I

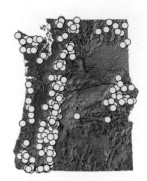

KEY FEATURES:
- Cespitose, with many slender, arching culms
- Several short, separated spikes
- Narrow, often glaucous leaves
- Perigynia distinctly beaked

DESCRIPTION: Habit: Cespitose. **Culms**: 14-66 cm long. **Leaves**: 0.9-2 (-2.3) mm wide, glaucous to green. **Inflorescences**: Spikes separated (or upper ones crowded), 4-7, greenish to brown, usually gynecandrous (or ♀), with 4-18 ♀ flowers, 5.5-9.8 mm long. **♀ scales**: Straw-colored to reddish brown, shorter than the perigynia. **Perigynia**: Ascending (the beaks sometimes spreading, but not the perigynia bodies), green to brown, sometimes reddish brown apically, 2.3-3.7 mm long, lacking swollen, pithy bases, with beak 0.4-1.1 (-1.3) mm long, toothless or with tiny teeth to 0.1 mm long. Stigmas 2. **Achenes**: Lenticular.

HABITAT AND DISTRIBUTION: Moist forest, moist meadows, bogs, seeps, shady streamsides; widely distributed but mainly in forested areas. AK to MT, S to CA and WY.

IDENTIFICATION TIPS: *Carex laeviculmis* is a densely cespitose plant with many slender leaves and many arching culms, each with several separated, short spikes. It is similar to *C. leptopoda* and its relatives, which grow in the same shaded, streamside habitats but in a slightly drier zone. *Carex leptopoda* and its relatives are coarser plants with wider leaves, and their lateral spikes and perigynium beaks both average longer. Their perigynia have slightly pithy bases. *Carex echinata* perigynia are strongly spreading to reflexed, spreading from the base but with the beaks not bending back, and they average longer, with more swollen, pithy bases. A small, esoteric distinction: *C. laeviculmis* perigynia have the attachment scar at the base; those of *C. echinata* have the scar on the dorsal surface near the base. Young or small-fruited *C. laeviculmis* are often confused with *C. brunnescens,* which has smaller perigynia with shorter beaks, and ♀ scales about as long as the perigynia but not concealing them. *Carex brunnescens* occurs in N WA and the WA Cascades.

COMMENTS: Surprisingly, this species is absent from the mountains of the northern Great Basin. A well-developed plant with dozens of arching culms would make a graceful addition to a shady garden.

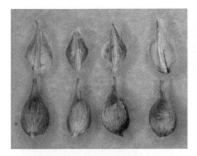

top left: pistillate scales,
  perigynia
top right and center left:
  inflorescences
*center right and bottom
  left*: habit

Carex lasiocarpa Ehrh.
Common name: Slender Woolly Sedge
Section: *Paludosae*   Key: C
Synonyms: *C. lasiocarpa* var. *americana* (Fern.) Hultén

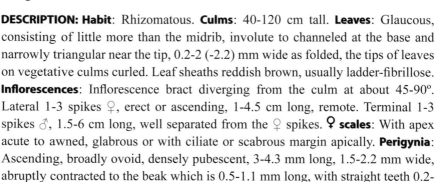

KEY FEATURES:
- Perigynia pubescent
- Leaf sheaths glabrous, fibrillose, and shiny, varnished reddish brown
- Leaves very narrow, triangular in cross section
- Bogs and fens

DESCRIPTION: Habit: Rhizomatous. **Culms**: 40-120 cm tall. **Leaves**: Glaucous, consisting of little more than the midrib, involute to channeled at the base and narrowly triangular near the tip, 0.2-2 (-2.2) mm wide as folded, the tips of leaves on vegetative culms curled. Leaf sheaths reddish brown, usually ladder-fibrillose. **Inflorescences**: Inflorescence bract diverging from the culm at about 45-90°. Lateral 1-3 spikes ♀, erect or ascending, 1-4.5 cm long, remote. Terminal 1-3 spikes ♂, 1.5-6 cm long, well separated from the ♀ spikes. ♀ **scales**: With apex acute to awned, glabrous or with ciliate or scabrous margin apically. **Perigynia**: Ascending, broadly ovoid, densely pubescent, 3-4.3 mm long, 1.5-2.2 mm wide, abruptly contracted to the beak which is 0.5-1.1 mm long, with straight teeth 0.2-0.7 mm long. Stigmas 3. **Achenes**: Trigonous.

HABITAT AND DISTRIBUTION: Fens, bogs, lakeshores, and wet meadows, sometimes forming floating mats, in deep, organic, acidic soils with low to moderate nutrient levels; in the Cascades of WA and OR; in the Puget Trough, Olympic Peninsula, and mts of N WA; and the mts of NE OR. AK to NF, S to CA (rare, Plumas Co.), and MT, MN, and VA, and Eurasia.

IDENTIFICATION TIPS: *Carex lasiocarpa* is a slender bog sedge with densely pubescent perigynia and narrow leaves more or less triangular (not flat) in cross section, with tips that curl. In a large stand, the divergent inflorescence bracts shining in the sun are distinctive. Very similar *C. pellita* is a shorter plant that grows in a greater diversity of wetlands and occupies shallower water when the two grow together. Its broader leaves are flat, V-shaped or M-shaped and do not curl at the tip. Both species frequently form large, sterile stands.

COMMENTS: *Carex lasiocarpa* is a community dominant in acidic fens and bogs. It has the ability to use organic phosphorus sources, valuable to plants growing in low-nutrient environments. Off-road vehicles can seriously damage the deep organic soils it grows in, harming populations. Genetic studies suggest that *C. lasiocarpa* may be the ancestor of the similar *C. pellita*.

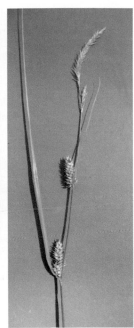

top left: perigynia
top middle: spike
top right: inflorescence
center left: habitat
bottom left: inflorescence

Carex leporina L.
Common name: Oval Broom Sedge
Section: *Ovales*   Key: J
Synonyms: *Carex ovalis* Gooden, *C. tracyi* Mackenzie

KEY FEATURES:
- Cespitose, with gynecandrous spikes & winged perigynia
- Inflorescence somewhat elongated
- Culms become stolons late in the season

DESCRIPTION: Habit: Cespitose. **Culms**: (24-) 35-85 cm tall at flowering. Vegetative shoots elongating late in the season and becoming stolons (developing shoots and roots at the nodes), sprawling over shorter vegetation. **Leaves**: 7-22 cm long, (1.5-) 2-3.5 (-4) mm wide. **Inflorescences**: 1.5-4 cm long, 10-15 mm wide, erect to flexuous, not head-like, rather open, the lower spikes somewhat separated. Spikes gynecandrous. ♀ **scales**: Straw-colored, brown or green, margins not white-hyaline but the tip often white-hyaline, with paler midstripe, 3.4-5 mm long, longer or shorter, wider or narrower than the perigynia, with tips acute to acuminate. **Perigynia**: Straw-colored or light brown, 3.4-4.7 (-5.2) mm long, 1.3-2.1 mm wide; wings 0.2-0.5 mm wide; with 3-9 conspicuous dorsal veins and 2-5 conspicuous ventral veins. Beak tip unwinged, brown, parallel-sided, and entire for the distal 0.3-0.5 mm (occasionally flat almost to the tip), with the very tip brown; (1.2-) 1.5-2 mm from achene top to beak tip. Stigmas 2. **Achenes**: Lenticular, 1.1-1.8 mm long, 0.9-1.2 mm wide, 0.4-0.5 mm thick.

HABITAT AND DISTRIBUTION: Moist meadows and wet prairies at low elevations; mostly W of the Cascades. Native from BC to CA, E to NV; introduced to eastern N America (WI to NF, S to TN and VA), Europe, Japan, Australia, and New Zealand.

IDENTIFICATION TIPS: *Carex leporina* is a loosely tufted sedge with a light brown, somewhat elongate inflorescence. The stolons, unusual in PNW *Ovales,* form late in the season and are rarely collected. Very similar *C. praticola* has ♀ scales with white margins. Its perigynia are veinless or weakly veined on the ventral surface, have longer beaks that are brown, parallel-sided, and wingless for the distal 0.4-1 mm and have the very tip narrowly white-edged.

COMMENTS: This ruderal species has been considered native to Europe and probably introduced to N America, but recent phylogenetic research suggests that *C. leporina* evolved in western N America and was introduced to Europe. Flip-flopping between names *C. leporina* and *C. ovalis* is due to confusion about which plant specimen should be considered the type for the name *C. leporina.*

top left: pistillate scales, perigynia
top right and bottom: inflorescences

Carex leporinella Mack.
Common name: Bog Hare Sedge
Section: *Ovales*   Key: J

KEY FEATURES:
- Cespitose, with gynecandrous spikes and winged perigynia
- Perigynia boat-shaped
- Spikes ascending, overlapping, in a narrow inflorescence
- Montane or alpine bogs and wet meadows

DESCRIPTION: Habit: Cespitose. **Culms**: 14-30 cm tall. **Leaves**: 10.5-15 cm long, 1-2.5 mm wide. **Inflorescences**: 1.4-3 cm long, 5-11 mm wide, spikes ascending to erect, overlapping, gynecandrous. ♀ **scales**: Straw-colored to reddish brown, generally with lighter midstripe, ovate, (2.5-) 3.5-6 mm long, usually longer and wider than the perigynium bodies (or the entire perigynia) and ± hiding them. **Perigynia**: Straw-colored to whitish or dull brown, opaque, lanceolate, 3.5-4.2 mm long, 1-1.2 mm wide, usually with (0-) 5-10 conspicuous dorsal veins, 0-6 conspicuous ventral veins. Wings very narrow, 0.05-0.2 mm wide, ciliate-serrulate, extending far up the beak and incurved toward the ventral side, the perigynia thus boat-shaped. Beak unwinged, brown, parallel-sided, and entire for the distal 0.3-0.6 mm, generally with a conspicuous white margin to the dorsal suture; 1.5-2.2 mm from beak tip to achene top. Stigmas 2. **Achenes**: Lenticular, 1.4-1.8 mm long, 0.7-1 mm wide, 0.3-0.4 mm thick.

HABITAT AND DISTRIBUTION: Marshes, bogs, and lake margins, in habitats that are soggy wet early in the season but may dry out later; mid-montane to alpine; S OR Cascades N to Mt Adams, WA; mts of NE OR and NE WA. WA to MT, S to CA and WY.

IDENTIFICATION TIPS: *Carex leporinella* is identified by the combination of short height, ascending spikes, ♀ scales that nearly conceal the perigynia; narrow, boat-shaped perigynia, and shallow-water habitat. It is often confused with *C. tahoensis* and *C. phaeocephala,* plants of dry subalpine to alpine sites. *Carex tahoensis* perigynia are similar in texture and venation to *C. leporinella* but larger. *Carex phaeocephala* perigynia are translucent, and veinless or nearly so on the ventral surface. *Carex abrupta* may have narrow boat-shaped perigynia, but it has a dense head and scales shorter than the perigynia.

COMMENTS: The wet habitats of PNW plants differ from the dry, rocky, alpine slopes where *C. leporinella* has been reported in CA. Most *Carex* species are microhabitat specialists and a single species would be unlikely to occur in such disparate habitats. More research is needed.

Carex leporinella

top left and middle: perigynia
top right: inflorescences
center left: habit
center right: inflorescence
bottom right: habitat (very wet meadow)

243

Carex leptalea Wahlenb.

Common name: Delicate Sedge, Jelly Bean Sedge
Section: *Leptocephalae*   Key: A
Synonyms: *C. leptalea* Wahl. ssp. *pacifica* Calder &
  Taylor

KEY FEATURES:
• Single spike
• Beakless perigynia shaped like rice grains

DESCRIPTION: Habit: Growing in clumps from short, pale rhizomes. **Culms**: 10-70 cm, very thin and lax. **Leaves**: Flat, soft, 0.4-1.3 mm wide. **Inflorescences**: One androgynous spike per culm, 0.4-1.5 (-1.8) cm long, the ♂ portion often short and inconspicuous but occasionally occupying most of the spike; ♀ part with 1-10 ± overlapping perigynia, lacking inflorescence bract. **♀ scales**: Usually about half as long as the perigynia, green or brown, sometimes awned. The lowest ♀ scales are sometimes prolonged as thin bristles. **Perigynia**: Green or straw-colored, 2.4-4.9 (-5.4) mm long, with 2 marginal ribs and many fine veins, with a spongy, stipe-like base, rounded at the top and virtually beakless. Stigmas 3. **Achenes**: Trigonous.

HABITAT AND DISTRIBUTION: Bogs, wet meadows, pond margins, other wetlands, in full sun or often in shade, often growing in moss mats, sometimes forming a lawn under willow thickets; W of the Cascade crest and in N WA and NE OR. Throughout N America, including Mexico and the Dominican Republic.

IDENTIFICATION TIPS: *Carex leptalea* is distinctive with its solitary spikes and almost beakless perigynia. Its foliage is soft, not bright green, sometimes forming a grass-like lawn under willows. *Carex tenuiflora,* rare in Okanogan Co., WA, is somewhat similar but has multiple small spikes clustered at the top of the culm and leaves 0.5-2 mm wide. Delicate though it is, it is more robust than *C. leptalea.*

COMMENTS: *Carex leptalea* is the most widely distributed *Carex* within N America. In the PNW it may be more overlooked than rare, partly because it is so delicate, partly because it is most common in dense willow thickets avoided by botanists. The taxonomy of *C. leptalea* is unsettled. Two subspecies could be recognized in our area. Widespread *Carex leptalea* ssp. *leptalea* is the more delicate plant with green to brownish ♀ scales and smaller (2.5-3.5 mm long) perigynia. *Carex leptalea* ssp. *pacifica* occurs from AK to Thurston Co., WA, W of the Cascade crest. It has brown-margined scales, lower ♀ scales with a distinct point or awn, and larger perigynia [(3-) 3.4-4.7 mm long]. We do not find this distinction clear and choose not to recognize these subspecies at the present time.

Carex leptalea

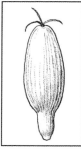

top left and middle:
  perigynia
top right and center left:
  inflorescences
bottom left: habit

Carex leptopoda Mack.

Common name: Slender-foot Sedge
Section: *Deweyanae*   Key: I
Synonyms: *C. deweyana* Schwein. ssp. *leptopoda*
   (Mack.) Calder & Taylor

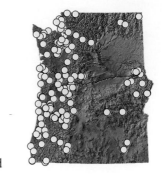

Key features:
- Loosely cespitose, with lax, grass-like leaves
- Moist forest
- Thin, membranous perigynia with short beaks and short teeth
- Ligules longer than wide

DESCRIPTION: Habit: Cespitose. **Culms**: 21-67 cm long, smooth or papillose (at 20X) at mid-length, the fragile papillae like tiny pegs projecting at right angles to the culm on its edges. **Leaves**: 2.4-5.9 mm wide, with ligule of uppermost culm leaf longer than wide, (2.5-) 3.4-6.8 (-7.4) mm long. **Inflorescences**: Spikes 5-7, usually longer than wide, green and white (maturing light brown), relatively smooth in outline, usually gynecandrous (or ♀); lowest spikes 8-15 mm long, usually with 11-23 perigynia; terminal spikes 9-12.7 mm long. ♀ **scales**: White (to straw-colored) with green midvein, usually awned, the awn to 0.7 mm long, the body shorter than the mature achene. **Perigynia**: 3.3-3.9 (-4.4) mm long, particularly thin-textured, with beak 0.9-1.5 (-1.7) mm long, 30-38% of perigynium length, with teeth to 0.3 mm long. Stigmas 2. **Achenes**: Lenticular. **Anthers:** 1.8-1.9 mm long.

HABITAT AND DISTRIBUTION: Very common in shaded to partially shaded riparian zones and moist forests, especially W of the Cascades, less common in mts E of the Cascades. BC to MT and CA.

IDENTIFICATION TIPS: This species grows in moist forest and is loosely cespitose with green, lax, grass-like leaves, separated spikes, and slender perigynia. It is one of the "identical triplets," along with *C. bolanderi* and *C. infirminervia.* All were formerly treated as subtaxa of *C. deweyana. Carex leptopoda* is the short-beaked, short-toothed one with pale green inflorescences, its perigynia seeming to swell out of the white ♀ scales, exposing the fragile, often torn, dorsal surface. *Carex bolanderi* has spikes that are more straw-colored and jagged, longer beaks with longer spreading teeth, and smooth or scabrous culms (at mid-length). *Carex infirminervia* has longer beaks. In N WA, compare to *C. deweyana* var. *deweyana,* which has shorter spikes, about as wide as long, larger perigynia and the ligules of upper leaves about as long as wide. In W OR and WA, *C. leptopoda* often occurs with *C. hendersonii,* which has wider leaves and larger, trigonous perigynia.

COMMENTS: This is the most common sedge of moist conifer forest in WA and NW OR. See discussion of *C. infirminervia* and *C. bolanderi.*

top left: perigynia
top right: inflorescences
center left: habit
bottom: habit with
  inflorescences

Carex limosa L.
Common name: Mud Sedge
Section: *Limosae*   Key: F

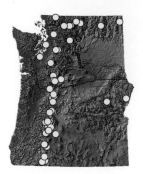

KEY FEATURES:
- Spiky-looking plant; with 1 or 2 thin, upright leaves and long inflorescence bract
- Spikes brown, lower ones pendent on long peduncles
- Bogs and wet pond margins

DESCRIPTION: Habit: Stoloniferous (and with inconspicuous vertical rhizomes), the shoots arising singly or in small groups. Young roots with yellow felty hairiness. **Culms**: 20-60 cm tall. **Leaves**: Few, 1-2.5 mm wide, glaucous, channeled or grooved, margins involute. **Inflorescences**: Lowest inflorescence bract 2-7 mm long, leaf-like. Lateral spikes ♀ or androgynous, 0.6-2 cm long, 4-8 mm wide. Terminal spike ♂, 0.7-3.5 cm long, 1.5-2.5 (-3) mm wide. ♀ **scales**: Yellowish brown to dark brown, 3-5.5 mm long, 2-3.4 mm wide, wider than the perigynium at the base, but often somewhat narrower than the perigynium above, as long as or longer than the perigynium, with apex obtuse to more or less acute, sometimes mucronate. **Perigynia**: Pale, green to straw-colored, papillose, broadly oval, 2.5-4 mm long, 1.8-2.6 mm wide, apex rounded, beak 0.1-0.5 mm long. Stigmas 3. **Achenes**: Trigonous.

HABITAT AND DISTRIBUTION: Mainly nutrient-poor fens and bogs, floating peat mats; wet, peaty meadows and pond margins; Puget Lowlands, mts of N WA, Cascades of WA and OR, and NE OR. Circumboreal; AK to NF, S to CA, UT, NE, and DE.

IDENTIFICATION TIPS: *Carex limosa* is a spiky little sedge putting up separated shoots, each with a couple of short leaves, the bract, and the narrow ♂ spike all pointing up and the short, broad ♀ spikes hanging down like little lanterns. The only similar species are *C. magellanica* and *C. pluriflora*, both rare in the PNW. *Carex magellanica* differs in its blue-green and blackish spikes and its pistillate scales that are narrower and longer than the perigynia, making the spikes look shaggy. *Carex pluriflora* differs in its beakless perigynia, usually blackish spikes, and fertile culms with remnants of dead leaves at the base.

COMMENTS: "Mud Sedge" is too mundane a name for this graceful little bog-dweller. Inconspicuous *C. limosa* may be a community dominant in its restricted habitat. Its buried, rhizome-like stolons are shallow, but its roots penetrate more deeply into its organic soil substrate than those of nearby sedges. *Carex limosa* is vulnerable to influxes of nutrients to its low-productivity habitats from sedimentation, faulty septic systems, and agricultural run-off. The added nutrients shift the competitive balance in favor of taller species.

top left: perigynia
top right: spike
center left: spike
bottom left:
    inflorescences, habitat
    (wet meadow)

Carex livida (Wahlenb.) Willd.
Common name: Pale Sedge
Section: *Paniceae*   Key: F

KEY FEATURES:
- Rhizomatous, with glaucous foliage
- Perigynia pale and papillose
- Bog habitat

DESCRIPTION: Habit: Rhizomatous, with slender rhizomes. **Culms**: 10-45 (-55) cm tall. **Leaves**: Glaucous, 4.5-40 cm long, channeled, 1-3.5 mm wide. Plant bases whitish or pale to medium brown. Lower leaves usually with blades. **Inflorescences**: Lateral spikes usually 1-2, ♀, 0.7-2.5 cm long, 4-5.5 mm wide. Terminal spike ♂, 1-3.3 cm long, 1.4-5 mm wide, with 16-100 flowers. ♀ **scales**: ± as long as or somewhat longer than the perigynia, pale to very dark brown with green midrib, obtuse. **Perigynia**: Widest in the middle and tapered to both ends, 3-4.8 (-5) mm long, 1.4-2.1 mm wide, light green or whitish, papillose. Stigmas 3. **Achenes**: Trigonous.

HABITAT AND DISTRIBUTION: Open bogs and fens with the water table at or close to the surface, usually with extensive peat deposits but sometimes on mineral substrates next to slow streams; along coast and in the Cascades of WA and OR and the Olympic Peninsula of WA. Extirpated from southernmost sites along the Pacific Coast in Mendocino Co., CA, and Lincoln Co., OR. Range interruptedly circumboreal; AK to NF, S to MA and OR; Eurasia, Panama, S America.

IDENTIFICATION TIPS: Seen from a distance, the grayish color of *C. livida* can give a bog a moldy appearance. It is a glaucous, narrow-leaved, rhizomatous sedge with pale, papillose, spindle-shaped perigynia. *Carex klamathensis* lives in serpentine fens in SW OR and NW CA and has flat leaves and obovate perigynia with beaks usually bent to the side. *Carex crawei* grows on limestone substrates and has green leaves and non-papillate perigynia speckled with reddish brown dots.

COMMENTS: In the NE U.S., *Carex livida* is confined to calcareous bogs with pH above 5.5, but in the PNW it occupies acidic bogs on non-calcareous substrates. This is a boreal species that has retreated northward with the glaciers, leaving isolated remnant populations behind. It is rare in most jurisdictions at the southern edge of its range. Management requires preventing changes in drainage patterns of its bog habitats. The few remaining coastal populations are at risk due to development and farming.

top left: pistillate
   scales, perigynia
top right: spikes
center left: habit
center middle:
   perigynium
bottom: habitat
   (*Sphagnum*
   bog with dark
   green *Equisetum*
   fluviatile and
   gray-green *C.*
   livida)

Carex longii Mack.
Common name: Green-and-White Sedge
Section: *Ovales*   Key: J

KEY FEATURES:
- Cespitose, with gynecandrous spikes and winged perigynia
- Wide, obovate perigynia
- Green leaf sheath fronts
- Inflorescence a line of distinct, green, somewhat pointed spikes

DESCRIPTION: Habit: Cespitose. **Culms**: 30-120 cm, the vegetative culms sometimes rooting at nodes. **Leaves**: 8-30 cm long, 2-4.5 mm wide; leaf sheath front green and veined except for a small white-hyaline triangle at the top. **Inflorescences**: 1-4.5 (-6) cm long, 5-14 mm wide, typically somewhat elongate, the spikes distinct but overlapping, gynecandrous, rounded or tapered to tip. ♀ **scales**: White-hyaline, aging silvery brown, with green center, 2.2-3.7 mm long, shorter and narrower than the perigynia, with apex obtuse. **Perigynia**: Green or light brown, 3-4.6 mm long, obovate 1.6-2.6 (-2.8) mm wide, the body widest above the middle; wings 0.5-0.8 mm wide, with 5 to many veins on each face. Beak tip winged and ciliate-serrulate almost to the tip; 1.4-2.2 mm from achene top to beak tip. Stigmas 2. **Achenes**: Lenticular, 1.3-1.7 mm long, 0.7-1 mm wide, 0.4-0.5 mm thick.

HABITAT AND DISTRIBUTION: Bogs and pond shores, on peaty or sandy soils, often where water levels fluctuate greatly, sometimes a weed in cranberry bogs; introduced to coastal WA and OR. Native WI to ME, S to TX and FL, Mexico, and Ecuador. Introduced to CA, Hawaii, New Zealand.

IDENTIFICATION TIPS: *Carex longii* is a cespitose species with a silvery, somewhat elongated inflorescence and green leaf sheath fronts. Its obovate perigynia are unusual. It has a strong superficial resemblance to *C. feta,* which lives in wet meadows, has spikes always rounded at tip, and has ovate perigynium bodies that are widest near or below middle.

COMMENTS: This species was recently introduced to the coastal PNW as a contaminant in cranberry vines brought from New England. How widely it will spread into native wetlands is unknown, but suitable habitat appears to exist. Sedges in *Ovales* (and a few other species) have true vegetative culms with some leaves originating at above-ground nodes. In the introduced *C. longii,* and *C. tribuloides,* as well as native *C. leporina,* vegetative culms are elongated with relatively conspicuous nodes. These culms may root at the nodes and send up new shoots, thus functioning as stolons.

Carex longii

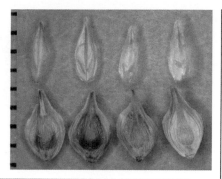

top left: pistillate scales,
   perigynia
top right: inflorescence
left: habit

Carex luzulina Olney

Common name: Woodrush Sedge
Section: *Aulocystis*   Key: F, (C)
Synonyms: *C. luzulina* var. *luzulina, C. luzulina* var.
   ablata (L. H. Bailey) F. Hermann, *C. ablata* L. H. Bailey

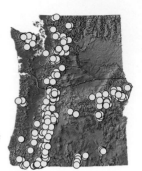

KEY FEATURES:
• Leaves short and broad, *Luzula*-like, clustered at base
• Perigynia green to purple, narrowly lanceolate
• Bogs and wet meadows, mainly in mountains

DESCRIPTION: Habit: Cespitose to short-rhizomatous. **Culms**: 15-90 cm tall, much longer than the leaves. **Leaves**: 3-9 mm wide, glabrous. Leaf sheath fronts whitish hyaline. **Inflorescences**: Lowest inflorescence bracts with ± inflated sheaths (5-) 10+ mm long. Lateral spikes crowded around the terminal one or separated, erect and nearly sessile or the lower ones drooping on long, slender peduncles, ♀ or androgynous, 1.2-3.2 cm long. Terminal spike ♂, 1.1-6.6 cm long. ♀ **scales**: Reddish brown to dark purplish with a light mid-stripe, sometimes white-edged, shorter than the perigynia, margins glabrous. **Perigynia**: Ascending to spreading, either mostly purple or green (with reddish or purple spots), variable in shape, with several inconspicuous veins, (3-)3.5-5.5 mm long, 0.9-1.6(-1.8) mm wide, usually glabrous, sometimes with sparse, spreading to appressed, usually long, soft, thin hairs, particularly on the beak, the body without flat margins, or rarely with flat margins less than half as wide as the achene, gradually tapering to the beak; distance from achene top to beak tip less than 2.5 mm. Stigmas 3. **Achenes**: Trigonous.

HABITAT AND DISTRIBUTION: Bogs and wet meadows, (100-) 3000-8000 feet elevation (to 10600 feet in WY); in mts and rarely near coast in the PNW. BC to WY, S to CA.

IDENTIFICATION TIPS: Somehow the broad, yellow-green leaves clustered at the base of the culm, together with the inflorescence bracts with ± inflated sheaths, the pointy perigynia, and the reddish to purplish ♀ scales, make *C. luzulina* unlike any other sedge in bogs of WA and OR. However, it's hard to point at any one consistent feature as identifying this species. Variable *C. luzulina* is confused with Californian *C. luzulifolia* and *C. fissuricola,* related species with perigynia with flat margins more than half as wide as the achene. In *C. fissuricola*, the perigynium faces, at least near the beak, have sparse, spreading-ascending, short, stiff bristles and its pistillate scales have hairs on the distal margins.

COMMENTS: *Carex luzulina* is a community dominant in bogs and wet meadows. It is variable in clustering of spikes, color and shape of scales, and color of perigynia. Two varieties (sometimes treated as species) have been recognized in the PNW, but the traits supposedly distinguishing them vary independently. We do not recognize them, but you might want to try.

top left: perigynia
top middle: inflated inflorescence bract sheath
top right and bottom left: inflorescences
bottom right: inflorescences on leaning culms

Carex lyngbyei Hornem.
Common name: Lyngbye's Sedge
Section: *Phacocystis*   Key: G

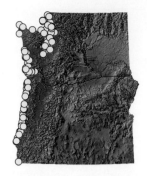

KEY FEATURES:
- Forms dense monocultures in salt marshes
- Spikes blunt based, dangling on curved stalks
- Perigynia hard, yellow-brown

DESCRIPTION: Habit: Rhizomatous, but the deep rhizomes often not collected; forming dense stands. **Culms**: 25-130 cm. Plant bases reddish brown. **Leaves**: Green, 3-8 mm wide, smooth-edged, dying in winter. Leaf sheath front hyaline, not spotted, not ladder-fibrillose. **Inflorescences**: Lowest inflorescence bract usually longer than the inflorescence. Lateral 2-4 spikes ♀ or androgynous, dangling on curved peduncles (but the spike itself does not bend), usually truncate at the base, 1.8-5 cm long, 5-7 mm wide. Terminal 2-3 spikes ♂. ♀ **scales**: Reddish brown to dark purplish brown, longer than the perigynia, acuminate. **Perigynia**: Yellowish brown with pale brown spots, ovate to obovate, hard, 5-7 veined on each face, 2.5-3.5 mm long, 1.6-2.5 mm wide, larger than the enclosed achene. Beak 0.1-0.3 mm long, not bidentate. Stigmas 2. **Achenes**: Lenticular, often with a dent in the side, like a dented beer can.

HABITAT AND DISTRIBUTION: Coastal salt marshes, and in ± fresh water along the lower Columbia River; AK to CA, also Greenland, Iceland, and Europe.

IDENTIFICATION TIPS: *Carex lyngbyei* forms extensive stands in coastal salt marshes. Its ♀ spikes are relatively short, with truncate bases, dangling at the ends of curved stalks. Similar *C. obnupta* and *C. aquatilis* var. *dives* grow in fresh to brackish water, sometimes adjacent to *C. lyngbyei* stands. Both have longer, usually sessile ♀ spikes with tapered bases. *Carex obnupta* has dark brown perigynia and evergreen leaves with tiny, sharp prickles on the margins.

COMMENTS: *Carex lyngbyei* foliage is an important food for migrating geese and nesting Tundra Swans, and coastal populations of Brown Bears in Alaska. Its decomposing leaves provide a rain of nutrients all winter long in estuaries, thus indirectly feeding young salmon. It is an important hay and silage crop in Iceland, with yields of 8-12 tons per acre. Leaves and rhizomes were used for weaving by Native Americans. In OR, *C. lyngbyei* has been displaced from some brackish marshes by invading Yellow Iris (*Iris pseudacorus*). Restoration projects establish plants from plugs, not seeds.

top left: pistillate scales, perigynia, achenes
top middle: leaf sheath fronts
top right: inflorescence
bottom: habit, habitat

Carex macrocephala Willd. ex Spreng.
Common name: Big-head Sedge
Section: *Macrocephalae*   Key: F, B

KEY FEATURES:
- Giant, spiky inflorescences
- Very large perigynia
- Coastal sand dunes

DESCRIPTION: Habit: Rhizomatous. **Culms**: 10-35 cm tall, sharply 3-angled, the culms serrulate near the inflorescence (on at least one angle). **Leaves**: 4-8 mm wide, longer than the culms but curving sideways or down, thus often not taller, serrulate on margins. **Inflorescences**: Each culm with a single dense head consisting of many densely aggregated spikes. Each head either ♂ or ♀, the plant thus appearing dioecious, although ♂ and ♀ shoots are connected by rhizomes. ♂ inflorescences 3.5-5 cm long and 7-15 mm wide. ♀ inflorescences 3.5-8 cm long, 2.5-5 cm wide, with lower spikes ± spreading. ♀ **scales**: With a short awn 1.2-4 mm long. **Perigynia**: 10-15 mm long, dark brown, shiny, spongy-thickened at the base. Margins with erose wings that curve inward over the ventral surface. Beak about as long as the body, with distance from the achene to the sharp tip 6.5-9 mm. Beak tip notched 0.7-1.5 mm. Stigmas 3. **Achenes**: Rounded-trigonous. **Anthers**: 2.5-5 mm long.

HABITAT AND DISTRIBUTION: Coastal sand dunes; AK to Coos Bay, OR. Also northern Japan, Russia; introduced to NJ.

IDENTIFICATION TIPS: With its big heads and its perigynia a centimeter or more long, *C. macrocephala* could only be confused with *C. kobomugi,* introduced to sandy soils, Portland, OR, to the coast of SW WA. *Carex kobomugi* has blunt angles on the stem, longer anthers, less strongly tapered, paler perigynia, and narrower flat margins.

COMMENTS: Once a community dominant with American Dune Grass (*Leymus mollis*) on the low dunes that were originally characteristic of the Pacific Coast, *C. macrocephala* has been extirpated from much of its former PNW range, mainly by sand dune stabilization by European Beach Grass (*Ammophila arenaria*). Large populations survive in limited areas (e.g., near Tillamook). Housing developments have destroyed some of its habitat. Some populations have been treated with herbicides because walking barefoot on the sharp perigynia is painful. A tincture of the seeds has been used as a stimulant in the Russian Far East. Both seeds and rhizomes may be edible for humans. *Carex macrocephala* plants are commercially available and would be a distinctive contribution to any coastal PNW garden with sandy soils, but might tend to spread.

top left: staminate inflorescences
top middle: perigynia
top right: pistillate inflorescence
bottom: habit

Carex macrochaeta C. A. Mey.
Common name: Long-awn Sedge
Section: *Limosae,* or perhaps *Scitae*   Key: F

KEY FEATURES:
- Clumps with dangling culms
- Dark, awned ♀ scales usually contrast with pale perigynia
- Wet rocks and cliffs, waterfall spray zones

DESCRIPTION: Habit: Loosely cespitose. Young roots with white, cream-colored, or yellowish felty hairiness. **Culms**: 20-50 (-70) cm long. Leaves: 2-5 mm wide, ± white papillate (at 15X) on the lower surface. **Inflorescences**: 3-5 spikes. Lateral ones ♀, 1-3 cm long, 6-8 mm wide, the lower spikes often dangling on moderately long peduncles but upper ones short-pedunculate to sessile. Terminal spike ♂, 1.2-3.2 cm long, 3-8 mm wide. **♀ scales**: Dark brown or black, often with a lighter midrib, narrower and longer than the perigynium (though the scale body itself may be shorter). ♀ scales awned, awns 1-3 mm long in the PNW (to 12 mm long in more northern sites). **Perigynia**: Narrowly elliptic, light green or sometimes marked with dark purplish color, 10-15 veined, 3.8-6.8 mm long, 1.7-2.3 mm wide, with beak 0.1-0.3 mm long. Stigmas 3. **Achenes**: Trigonous and filling only about 60% of the perigynium body.

HABITAT AND DISTRIBUTION: In the PNW, in the spray zone of waterfalls or on seepy, N-facing cliffs, always at cool sites; known from the Columbia Gorge (Skamania Co., WA, and Multnomah Co., OR), Saddle Mountain, Clatsop Co., OR, and the Willapa Hills of SW WA. To the N, diverse wet sites including shorelines, bogs, and marshes, from seaside to subalpine zone at about 3000 feet elevation; AK to S BC, disjunct in S WA and N OR; also coastal NE Asia.

IDENTIFICATION TIPS: *Carex macrochaeta* forms dense clumps of leaves that hang down from its cliff-side perch. *Carex scirpoidea* var. *stenochlaena* produces similar tufts in similar habitats, but has only one spike per culm. *Carex limosa, C. pluriflora,* and *C. magellanica* grow on more level sites and the first two lack ♀ scale awns.

COMMENTS: North of the PNW, *C. macrochaeta* is often a community dominant. Young shoots are high in protein and are important in the diet of species as diverse as Aleutian Canada Goose, Mountain Goat, and Brown Bear. In fact, some bear families move in response to seasonal changes in development of *C. macrochaeta* shoots. *Carex macrochaeta* has been retreating N with the glaciers, leaving small, isolated populations behind. It is rare along the southern border of its range.

Carex macrochaeta

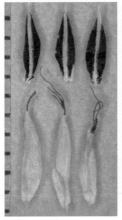

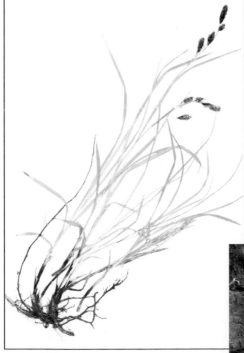

top left: pistillate scales, perigynia
top right: inflorescence
center left: habit
bottom right: habit, habitat

Carex magellanica Lam. ssp. irrigua
(Wahlenb.) Hiitonen
Common name: Boreal Bog Sedge
Section: *Limosae*   Key: F
Synonyms: *C. paupercula* Michx.

KEY FEATURES:
- Bicolored spikes with pale perigynia and narrow dark scales
- Terminal ♂ spike erect, ♀ spikes mostly dangling
- Bogs

DESCRIPTION: Habit: Rhizomatous, the shoots usually arising in small clusters, the plant sometimes growing as a clump. Young roots with yellow felty hairs. **Culms**: 10-80 cm tall. Leaves: 1-4 mm wide, green or glaucous, flat, the margins revolute. **Inflorescences**: Lowest inflorescence bract 2-10 cm long, leaf-like, with a short sheath < 4 mm long. Lateral spikes ♀ or gynecandrous, 0.6-2.2 cm long, 4-8 mm wide. Terminal spike ♂ or occasionally gynecandrous, 0.7-2 cm long, 1-4 mm wide. **♀ scales**: Brown to blackish, 2.8-7 mm long, 1.2-2 mm wide, narrower and longer than the perigynium, with apex acute to acuminate and often with an awn to 3 mm long, giving the spike a somewhat shaggy appearance. **Perigynia**: Pale, green to straw-colored, papillose, broadly oval, 2.5-3.6 mm long, 1.8-2.5 mm wide, apex rounded, beak 0-0.2 mm long. Stigmas 3. **Achenes**: Trigonous.

HABITAT AND DISTRIBUTION: Mainly *Sphagnum* bogs, also wooded wetlands, wet meadows, and lake shores; northern tier of counties in WA. Circumboreal, AK to NF, S to WA, MN, and CT, and to ID and CO in Rocky Mts., also Eurasia.

IDENTIFICATION TIPS: *Carex magellanica* is a spiky-looking sedge like *C. limosa,* but its spikes have contrasting blue green perigynia and dark brown pistillate scales. Its foliage is more glaucous, too. *Carex limosa* has grooved leaves and ♀ scales that are shorter, broader, and brown, lack awns and are only slightly longer than the perigynia. *Carex pluriflora* has wide, blackish ♀ scales that nearly hide the perigynia, and its lowest inflorescence bract is bristle-like, only 0.5-2 cm long.

COMMENTS: Like several other sedges shifting their ranges northward in response to climate change since the last glacial maximum, *C. magellanica* ssp. *irrigua* has a discontinuous distribution at the southern edge of its range. It is rare in jurisdictions as diverse as CO, WI, Ireland, and Poland. This species has a bipolar distribution, with *C. magellanica* ssp. *magellanica* growing in temperate S America.

Carex magellanica ssp. *irrigua*

top left: pistillate scales, perigynia
top right and bottom: inflorescences

Carex media R. Br. ex Richardson

Common name: Scandinavian Sedge
Section: *Racemosae*   Key: F
Synonyms: *C. norvegica* Retz. ssp. *inferalpina* (Wahl.)
  Hultén

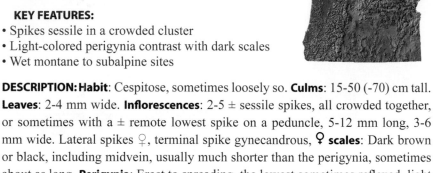

KEY FEATURES:
- Spikes sessile in a crowded cluster
- Light-colored perigynia contrast with dark scales
- Wet montane to subalpine sites

DESCRIPTION: Habit: Cespitose, sometimes loosely so. **Culms**: 15-50 (-70) cm tall. **Leaves**: 2-4 mm wide. **Inflorescences**: 2-5 ± sessile spikes, all crowded together, or sometimes with a ± remote lowest spike on a peduncle, 5-12 mm long, 3-6 mm wide. Lateral spikes ♀, terminal spike gynecandrous, ♀ **scales**: Dark brown or black, including midvein, usually much shorter than the perigynia, sometimes about as long. **Perigynia**: Erect to spreading, the lowest sometimes reflexed, light green becoming tan, elliptic or ovate, 2.5-3.5 mm long, 0.9-1.5 mm wide, with beak 0.3-0.4 mm long and darker than the body. Stigmas 3. **Achenes**: Trigonous, nearly filling the perigynium bodies.

HABITAT AND DISTRIBUTION: Montane to subalpine meadows, streambanks, seepage areas, bog margins; Okanogan Co., WA, and Wallowa Mts, OR. AK to NF, S to CA (Mono Co.), MT, IA, and ME.

IDENTIFICATION TIPS: *Carex media* is a cute montane sedge, with small, crowded, mostly sessile spikes having green perigynia that contrast with dark scales. *Carex pelocarpa* is somewhat similar, but it has dark inflorescences with ♀ scales about as long as the obovate to circular, relatively long perigynia. It grows in drier habitats above timberline. *Carex albonigra* and *C. atrosquama* have crowded inflorescences like *C. media*, but their spikes have short peduncles. They also differ from *C. media* in perigynium color, which is dark brown in *C. albonigra* and dark golden in *C. atrosquama*.

COMMENTS: Like many PNW sedges, *C. media* has been retreating north since the end of the last ice age, leaving small, isolated populations behind. It is rare along the southern border of its range. It is closely related to and sometimes considered to be the same species as Eurasian *C. norvegica*.

top left: perigynia
top right: inflorescences
center left: habit
bottom: habitat

Carex mendocinensis Olney ex W. Boott
Common name: Mendocino Sedge
Section: *Hymenochlaenae* (or *Porocystis*?)   Key: F

KEY FEATURES:
- Dense mounds of yellow-green foliage
- Narrow, rather long ♀ spikes
- Seeps and ditches, usually on serpentine soils

DESCRIPTION: Habit: Densely cespitose. **Culms**: 25-80 cm tall. **Leaves**: 2.5 –5.5 mm wide, flat, usually glabrous (sometimes sparsely hairy on one or both sides especially near the base). Lowest leaf sheaths reddish, bladeless, and glabrous (or with a few rough hairs). Upper sheaths grade from dark red to green on back. Leaf sheath fronts tan-hyaline, red-dotted, glabrous or sometimes pubescent at the top. **Inflorescences**: Inflorescence bract with a well developed sheath 8-40 mm long. Lateral spikes 3-5, ♀, cylindric, 1-6 cm long, 2.5-6 mm wide, each with a bract that has a well developed sheath. Terminal spike ♂ (rarely gynecandrous), sessile or nearly so, 1.8-3.8 cm long. **♀ scales**: ± brown to gold with white margins and green midrib, broadly ovate, shorter than the mature perigynia, with ciliate margins but otherwise glabrous. **Perigynia**: Glabrous or with sparse hairs near the tip, green to brown, more or less spotted with red, usually reddish at base, oblong-lanceolate, with 12-15 fine veins plus 2 marginal ribs, 2.7-5 mm long, 0.8-1.8 mm wide, with beak 0.3-1 mm long. Stigmas 3. **Achenes**: Trigonous.

HABITAT AND DISTRIBUTION: Seeps, small streams, and wet meadows, typically on serpentine substrates but sometimes on somewhat disturbed non-serpentine sites, at low to moderate elevations; S Lane Co., OR, S to NW CA.

IDENTIFICATION TIPS: There is no other sedge quite like this one, with its dense mound of often yellowish green leaves, slender ♀ spikes, usually ciliate margins of the inflorescence bract sheath and the ♀ scales, and the small perigynia. The few hairs sometimes observed on leaves and perigynia might cause confusion with *C. gynodynama,* which has wider, hairier leaves and densely hairy perigynia that are packed more densely into the spike, making the spikes wider.

COMMENTS: *Carex mendocinensis* is common in serpentine seeps in SW OR, but scattered populations occur on non-serpentine substrates as far N as Douglas and Lane cos. Some of these populations may have been introduced along logging roads. *Carex gynodynama* and *C. mendocinensis* hybridize; the hybrids are sterile. *Carex mendocinensis* is planted ornamentally for the attractive, dense tufts of leaves.

Carex mendocinensis

top left: perigynia
top right: inflorescence
center: habit
bottom left: habit, habitat

Carex mertensii J. D. Prescott ex Bong.
Common name: Mertens' Sedge
Section: *Racemosae*   Key: F

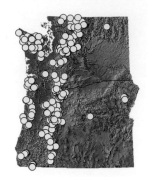

KEY FEATURES:
- Gracefully drooping inflorescence
- Fat, cylindrical spikes
- Perigynia flat, large, almost circular

DESCRIPTION: Habit: Cespitose. **Culms**: 30-100 cm tall. **Leaves**: 4-8 mm wide. **Inflorescence**: 4-9 spikes, each 1-4 cm long, the lateral spikes ♀ (or with a few ♂ flowers at base), the terminal spike gynecandrous. **♀ scales**: Brown to dark purplish black, shorter and much narrower than the perigynium, tip obtuse, mucronate. **Perigynia**: Ovate or orbicular, flattened except over the tiny achene, green to brown, 4-5 mm long, 2.5-3.5 mm wide, much longer and wider than the achene, with a tiny beak 0.2-0.4 mm long. Stigmas 3. **Achenes**: Trigonous, filling the lower half or less of the perigynium body.

HABITAT AND DISTRIBUTION: Mesic to somewhat dry open forest, meadows, roadsides, stream banks, the pumice plains formed following volcanic eruptions, and other disturbed places at low to moderately high (mostly 1500 - 6000 feet) elevations in the mountains, usually in partial shade; concentrated in the Cascades, Olympics, and Coast Range of OR and WA, and scattered in mts E of the Cascades. AK to SK, S to CA and MT, Asia.

IDENTIFICATION TIPS: Distinctive *C. mertensii* has thick, dangling, sausage-shaped spikes and is common on roadsides and other disturbed sites in mesic montane forest. It could be confused with *C. heteroneura* because its inflorescence may droop when fully mature. However, that species lives in subalpine to alpine habitats and has denser inflorescences, elliptic to obovate perigynia, and proportionately wider ♀ scales.

COMMENTS: *Carex mertensii* is independent of mycorrhizal fungi and has efficient uptake of nutrients. Therefore it can be an efficient colonizer of heavily disturbed, mesic sites. Its wide, flat perigynia can tumble long distances in windy sites, dispersing the seeds far from the original population. *Carex mertensii* has been recommended for roadside plantings because it establishes well in such sites and helps prevent soil erosion. The graceful inflorescences make the species an excellent candidate for the garden.

Carex mertensii

top left: perigynia
top right: inflorescence
center right: habit
bottom left: habitat (Mount
    St. Helens 10 years
    after eruption)

Carex micropoda C. A. Mey.

Common name: Timberline Sedge
Section: *Dornera*   Key: A
Synonyms: *C. pyrenaica* Wahl. ssp. *micropoda* (C. A.
 Meyer) Hultén, *C. crandallii* Gandoger

KEY FEATURES:
• Single spike with deciduous scales
• Densely cespitose with narrow, involute leaves
• High elevations in mountains

DESCRIPTION: Habit: Cespitose. **Culms**: 5-30 (-40) cm tall. **Leaves**: Involute, 0.25-1.5 mm wide (occasionally flat and to 2 mm wide, but always with involute leaves on the same plant). **Inflorescences**: One androgynous spike per culm, 0.7-2 cm long, 4-7 mm wide, lacking inflorescence bract. ♀ **scales**: Lanceolate, dark reddish brown to black, deciduous. **Perigynia**: Ascending when young, later spreading or even reflexed, lanceolate to elliptic, with a stipe at the base, dark brown or blackish distally, paler in lower half, glossy, 3-4 (-5) mm long, 1-1.3 mm wide, veinless but with 2 marginal ribs, tapered to a short, poorly defined, dark brown or black beak. Stigmas 2 or 3. **Achenes**: Lenticular or trigonous.

HABITAT AND DISTRIBUTION: Moist alpine meadows and snowmelt basins; mts of WA, Wallowa Mts. in OR. AK to AB, S to CA and CO, also Japan and E Russia.

IDENTIFICATION TIPS: *Carex micropoda* is a cespitose, narrow-leaved, single-spiked sedge of high elevations. With age, its perigynia spread and may even point downwards. Inflorescences of *C. nigricans* are very similar, but *C. nigricans* is rhizomatous and its leaves are flat and wider. *Carex circinata* is similar vegetatively, but its perigynia are very narrow.

COMMENTS: *Carex micropoda* grows mainly where heavy snow cover protects it from extreme cold, prevents growth of taller plants, and provides moisture during the growing season. It is an important food for pikas. Taxonomy of this complex is unsettled. European *C. pyrenaica* has 3 stigmas and wider perigynia. *Carex micropoda* in E Asia and NW N America has 2 stigmas and lanceolate perigynia. *Carex micropodioides* in N Asia has variable stigma number and perigynia with intermediate shape. PNW plants seem to be *C. micropoda* although they have variable stigma numbers. *"Carex crandallii"* of the Rockies and intermountain area has mostly 3 stigmas, with other variation overlapping *C. pyrenaica* and *C. micropoda*. Related *C. pyrenaica* var. *cephalotes* = *C. cephalotes*, with variable stigma number, grows in New Zealand and Australia. Expect nomenclatural changes.

Carex micropoda

top left: perigynia
top right: inflorescences
bottom: habit, habitat

Carex microptera Mack.

Common name: Small-wing Sedge
Section: *Ovales*   Key: J
Synonyms: *C. festivella* Mackenzie;, *C. limnophila*
    F. J. Hermann, *C. macloviana* d'Urv. var. *microptera*
    (Mackenzie) Boivin

KEY FEATURES:
* Cespitose, with gynecandrous spikes and winged
  perigynia
* Compact, two-toned head with fine texture due to
  many flat perigynia
* Common in mesic openings in mountains

DESCRIPTION: Habit: Cespitose. **Culms**: 20-110 cm. **Leaves**: 10-50 cm long, 2-5 mm wide. **Inflorescences**: (0.8-) 1-2.6 cm long, 9.5-17.5 mm wide, dense and head-like, usually two-toned because of contrast between the lighter color of perigynia and dark ♀ scales (but sometimes all dark), with a fine texture because the numerous perigynia are thin and appressed. Spikes gynecandrous (rarely ♀). ♀ **scales**: Generally brown or purplish with a paler midstripe, 2.4-3.5 mm long, shorter and narrower than the perigynia. **Perigynia**: Green or straw-colored but with dark beak tips, flat except over the achene, (2.8-) 3.4-4.5 (-5.2) mm long, 1.1-2.4 mm wide, wing 0.2-0.5 mm wide; with 3-8+ dorsal veins and 0-8 ventral veins. Beak tip unwinged, brown, and parallel-sided for the distal 0.6-1.3 mm; 1.5-2.5 (-2.8) mm from achene top to beak tip. Stigmas 2. **Achenes**: Lenticular, 1.1-1.6 mm long, 0.7-1.2 mm wide, 0.3-0.5 mm thick.

HABITAT AND DISTRIBUTION: Common in mesic forest openings and grasslands at moderate to high elevations in the mts; mainly in and E of the Cascades, in SW OR, and the Olympic Mts., WA. YT to NU and MB, S to CA, NM, and SD.

IDENTIFICATION TIPS: Cespitose *Carex* with winged perigynia, gynecandrous spikes, and a single, dense head are likely *C. microptera* or *C. pachystachya*. *Carex microptera* has fine-textured inflorescences and perigynia that are flat except over the small achene. In *C. pachystachya,* the fewer, thicker, planoconvex perigynia produce a coarser-textured inflorescence. It lives in moister habitats. Both species occur at high elevations, but *C. pachystachya* is also common in lowlands where *C. microptera* does not grow. In the alpine zone, *C. microptera* is replaced by *C. haydeniana,* with longer perigynia and shorter culms.

COMMENTS: *Carex microptera* is common and variable. Small-headed extremes have been segregated as different species. *Carex microptera* is used in habitat restoration projects. It tolerates soils with high heavy-metal content, such as mine tailings, and accumulates lead in its foliage.

top left: perigynia
top right: inflorescences
bottom: habit, habitat

Carex multicaulis L. H. Bailey
Common name: Many-stem Sedge
Section: *Firmiculmes*   Key: A

KEY FEATURES:
- Single spike with large perigynia
- Rush-like appearance
- Dry, open forest, often on serpentine soils

DESCRIPTION: Habit: Densely cespitose. **Culms**: 14-50 cm long, round in cross section. **Leaves**: Involute, 0.8-1.5 mm wide, usually much shorter than the culms. **Inflorescences**: One spike per culm, lacking inflorescence bracts; terminal part ♂, (0.8-) 1-2.5 cm long. ♀ **scales**: Green to brown, the lowest leaf-like, awned, and much longer than the perigynia, the upper ones reduced. **Perigynia**: Light green to light brown, obovate, 4.7-7.2 mm long, 2.3-3.1 mm wide, almost beakless. Rachilla present. Stigmas 3. **Achenes**: Trigonous.

HABITAT AND DISTRIBUTION: Dry slopes or seasonally moist areas in open conifer, oak, or mixed forest or in chaparral, often on serpentine substrates, at low to moderate elevations, Lane Co., OR, SW OR to CA.

IDENTIFICATION TIPS: At a glance, an observer could easily mistake this plant as a rush with its stiff, densely clumped culms and short, inconspicuous leaves. Although its perigynia are very similar to those of *C. geyeri,* which also occurs at scattered locations in SW OR, the two species look nothing alike. *Carex geyeri* is loosely cespitose to short-rhizomatous, with lots of flat leaves. Its culms are triangular in cross section, and its lower ♀ scales are short and not at all leaf-like. *Carex multicaulis* is more likely to be confused with a cespitose rush, unless the inflorescences are seen. Even then, a second look is needed. Although rush flowers are small and brown, they look much like conventional, unreduced flowers; they have 6 brown petals. *Carex* flowers are almost unrecognizable as flowers, hidden in the perigynium and lacking any petals. Rush fruits are capsules with many tiny seeds; *Carex* fruits contain a single, larger seed.

COMMENTS: At higher elevations, *C. multicaulis* becomes a relatively short plant with few culms or leaves. It is hard to recognize these plants as the same species as the robust plants of lower elevations. The relationship between these two forms is being researched.

top left: perigynia
top right: inflorescences
bottom: habit, habitat

Carex nardina Fries

Common name: Spikenard Sedge
Section: *Nardinae*   Key: A
Synonyms: *C. hepburnii* Boott, *C. nardina* var. *hepburnii* (Boott) Kük.

KEY FEATURES:
- Leaf blades very narrow, fescue-like
- Dense, persistent, fibrous leaf sheaths
- Perigynia glabrous except for serrulate beak margins
- Dry, open, high elevation places

DESCRIPTION: Habit: Densely cespitose. **Leaves**: Very narrow and wiry, to 0.5 mm wide, with dense, persistent, yellow-brown sheaths. **Inflorescences**: Spike single, androgynous, to about 1.5 cm long, with 5-15 perigynia, lacking inflorescence bracts. ♀ **scales**: Reddish brown or yellowish with broad brownish or white margins, about as wide as the perigynia. **Perigynia**: Glabrous but with serrulate distal margins, elliptic or lanceolate to oblong-obovate, green or light brown at maturity, planoconvex, with flattened margins distally, 3.5-4.5 mm long, 1.2-2 mm wide, beakless or with beak to 0.8 mm long, brown or hyaline. Rachilla well developed. Stigmas 2 or 3. **Achenes**: Lenticular or trigonous.

HABITAT AND DISTRIBUTION: Dry alpine ridges and slopes, abrasion plateaus, dry tundra, heaths, and fellfields; Cascades and Olympic Mts. of WA, mts of NE OR, one record from S Cascades in Douglas Co., OR. AK to LB, S to CO and NV, Eurasia.

IDENTIFICATION TIPS: *Carex nardina* is a densely cespitose, fine-leaved "bunch-grass" sedge of high elevations, with yellow-brown, persistent leaf sheaths. It is very similar to *C. filifolia,* which occasionally grows in subalpine areas and has pubescent perigynia that are more turgid. (Aborted *C. filifolia* flowers produce relatively flat perigynia that are difficult to distinguish from those of *C. nardina.*) *Kobresia myosuroides* grows in similar habitats and sterile plants have similar very narrow leaves and persistent, fibrous leaf sheaths. It has open perigynia which contain both a ♀ flower and a small ♂ flower.

COMMENTS: *Carex nardina* can be divided into two species in Europe, where slightly different plants live in different areas. Both morphologies can be found in N America but they seem not to be geographically segregated. A great deal has been written about this pattern of variation, but the situation remains anything but clear. Genetic research might help define the two taxa or show definitively that only one is present in N America.

top left: perigynia
top right:
   inflorescence
center left: habitat
   (with Fred
   Weinmann)
bottom left: habit,
   habitat
bottom right: habit

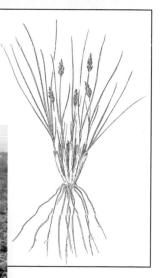

Carex nebrascensis Dewey

Common name: Nebraska Sedge
Section: *Phacocystis* Key: G

KEY FEATURES:

- Rhizomatous, with wide, usually glaucous leaves
- Perigynia strongly 5-9 veined
- Common E of the Cascades, usually in alkaline wetlands

DESCRIPTION: Habit: Rhizomatous, but the deep rhizomes not often collected. **Culms**: 20-90 cm. Plant bases brown or reddish brown. **Leaves**: Usually glaucous, (3-) 4-12 mm wide. Leaf sheath fronts hyaline, not ladder fibrillose, usually not spotted. **Inflorescences**: Lowest inflorescence bract ± equal to the inflorescence. Lateral 2-4 spikes ♀ (or androgynous), erect, blunt or only slightly tapering at base, 3-5.5 cm long, 5-8 mm wide. Terminal 1-3 spikes ♂. ♀ **scales**: reddish brown to dark purplish brown, acute to awned, the awn, if present, up to 0.5 mm long and sometimes scabrous. **Perigynia**: Brown with reddish brown spots, elliptic to ovate to obovate, tough, strongly 5-9 veined on each face at maturity, 2.6-4 mm long, 1.6-2.5 mm wide, with "sloping shoulders." Beak 0.3-0.6 mm, minutely bidentate. Stigmas 2. **Achenes**: Lenticular.

HABITAT AND DISTRIBUTION: Common in small, sunny riparian areas and other wetlands in otherwise very dry habitats, often in alkaline sites; mostly E of the crest of the Cascade Range, and in Jackson Co., OR. The few populations W of the Cascades are probably introduced. Widespread at low to middle elevations. WA to AB and ND, S to CA, NM, and KS. Introduced to MO, IL, and Japan.

IDENTIFICATION TIPS: This is a common, rhizomatous, broad-leaved, usually glaucous, lower-elevation sedge with strong perigynium veins. Some traits traditionally used to distinguish it from relatives (leaves W-shaped vs. V-shaped in cross section, or blue vs. green) do not work. *Carex aquatilis,* growing in ± acidic wetlands, differs in having veinless mature perigynia with rounded shoulders. *Carex angustata* has ladder-fibrillose leaf sheath fronts and mature perigynia with 1-3 weak veins on each face.

COMMENTS: *Carex nebrascensis* is an important forage and soil-binding species, common in non-acidic wetlands. Its habitats may experience moderately large annual fluctuations in water table. It can increase under grazing as competing vegetation is removed, and may persist by rhizomes in overgrazed wetlands. It is used for restoration of degraded wetlands, and for roadside plantings. Stunted relict plants may persist on terraces near streams that have been downcut 6 or 8 feet. Seeds disperse on vehicles, producing disjunct populations.

top left: perigynia   *top right and center left*: inflorescences
bottom left: glaucous, rhizomatous *C. nebrascensis* in streamside habitat
bottom right: stem bases

Carex nervina L. H. Bailey
Common name: Sierra Sedge
Section: *Vulpinae* Key: H

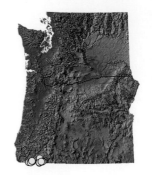

KEY FEATURES:
- Leaf sheath mouth with thickened rim
- Perigynia with swollen, pithy bases
- Moist meadows, seeps, SW OR

DESCRIPTION: Habit: Cespitose, forming a thick tuft of gracefully arching leaves. **Culms**: 60 to 70 cm tall, without persisting basal sheaths of previous years, somewhat winged, weak, bending down in fruit. **Leaves**: Green, 5 mm wide, arising at different heights so that the sheaths are mostly visible. Lower leaf sheaths herbaceous and veined, not splitting, longer than their reduced blades. Distal leaf sheaths hyaline, faintly veined, not cross-corrugated, with pale, thickened concave mouth. **Inflorescences**: Dense heads 1.5-2.2 cm long, (6-) 9-12 mm wide, with androgynous spikes (or some all ♀). ♀ **scales**: Brown, about as long as the perigynia. **Perigynia**: Lance-triangular, shaped somewhat like a bass viol, widest near the base, brown, to 4.5 mm long, about 2 mm wide, strongly veined, with about 15 veins on the dorsal surface and 7-12 veins on the ventral surface; the base only somewhat pithy and swollen, cordate, and with a stipe to 0.2 mm long; the beak poorly defined, to 2 mm long, with smooth to serrulate margins. Stigmas 2. **Achenes**: Lenticular.

HABITAT AND DISTRIBUTION: Wet meadows, seeps, and streamsides in the mountains, extreme southern Josephine and Jackson cos., OR; also NW CA and the Sierra Nevada of CA and NV.

IDENTIFICATION TIPS: *Carex nervina* produces graceful mounds of leaves. The tall but weak culms bend and look broken when mature. The thickened rim of the leaf sheath front is unique but sometimes hard to see. *Carex neurophora* is similar but has cross-corrugated leaf sheath fronts, less graceful foliage, and stronger culms that rarely bend. It lacks the thickened rim at the leaf sheath mouth.

COMMENTS: This plant is at the northern edge of its range in the Siskiyou Mts. of extreme southern OR. It usually grows in shade or partial shade, but established plants can survive if the canopy is removed.

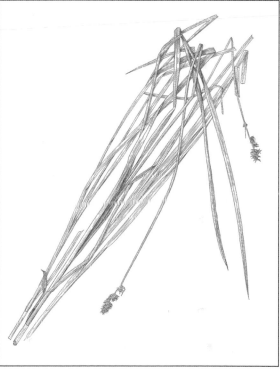

top left: perigynia   *top right*: inflorescences
bottom left: leaf sheath front with white, thickened rim at mouth
bottom right: habit

Carex neurophora Mack.
Common name: Veined Sedge
Section: *Vulpinae*   Key: H

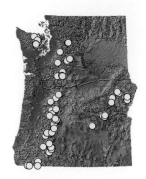

KEY FEATURES:
- Leaf sheath fronts cross-corrugated
- Perigynia with swollen, pithy bases
- Moist meadows, seeps.

DESCRIPTION: Habit: Loosely cespitose and often forming small mounds. **Culms**: 35 to 60 cm tall, lacking persistent basal sheaths of previous years. **Leaves**: Green, 5 mm wide, arising at different heights so that the sheaths are mostly visible. Sheaths of lower leaves herbaceous and veined, not splitting, longer than their reduced blades. Sheaths of upper leaves hyaline, faintly veined, usually cross-corrugated, with convex mouth. **Inflorescences**: A dense head 1-2.5 cm long, 6-15 mm wide, with androgynous spikes (or some all ♀). ♀ **scales**: Brown, shorter than or about equal to the perigynia. **Perigynia**: Lance-triangular, shaped somewhat like a bass viol, widest near the base, brown, to 2.9-3.8 mm long, about 1-1.5 mm wide, strongly veined, with 9-11 veins on the dorsal surface, 5-7 on the ventral surface, the base pithy and somewhat swollen, the beak poorly defined, to 2 mm long, with serrulate margins. Stigmas 2. **Achenes**: Lenticular.

HABITAT AND DISTRIBUTION: Wet meadows, seeps, and streamsides in the mountains from moderate to high elevations but not alpine; in the Cascades, Olympics, and mts of NE and SW OR. WA to MT, S to OR and CO.

IDENTIFICATION TIPS: *Carex neurophora* is a relatively tall, cespitose sedge with veined perigynia with swollen, pithy bases. Its leaf sheath fronts are usually cross-corrugated. It is most often confused with *C. jonesii*, which has leaves all clustered at the base and usually smooth leaf sheath fronts. Where the two grow together, *C. neurophora* seems a little taller, with denser shoots. The superficially similar *C. nervina* is found only in extreme SW OR and in NW CA. Its leaf sheath front has a thickened, white, concave mouth and its weak culms bend over. Its inflorescences are somewhat more elongated. Superficially similar sedges in section *Ovales* have tight, dark heads, but the *Ovales* have gynecandrous spikes, winged perigynia, and smooth leaf sheath fronts.

COMMENTS: *Carex neurophora* has a spotty distribution and is often missing from apparently suitable habitat. Its presence can be considered an indicator of good range conditions.

Carex neurophora

top left: perigynia
*top right and
    center right*:
    inflorescences
center left: cross-
    corrugated leaf
    sheath fronts
bottom left: habitat

Carex nigricans C. A. Mey.
Common name: Black Alpine Sedge
Section: *Dornera*   Key: A

KEY FEATURES:
• Single spike
• Perigynia reflexed when mature
• Snowmelt basins in mountains

DESCRIPTION: Habit: Rhizomatous and turf-forming, or loosely cespitose if growing among rocks. **Culms**: 5-30 cm tall. **Leaves**: Flat, (1.5-) 2-4 mm wide. **Inflorescences**: One androgynous spike per culm, (0.7-) 1-2 cm long, 6-10 mm wide, lacking inflorescence bract. ♀ **scales**: Lanceolate, dark reddish brown to black, deciduous before the perigynia mature. **Perigynia**: Ascending when young, then becoming spreading to reflexed, lanceolate to elliptic, with a stipe at the base, dark brown or blackish distally, paler in lower half, glossy, (3.0-) 3.2-4.6 (-5) mm long, 1-1.2 mm wide, veinless, tapered to the dark brown or black beak which is about as long as the body. Stigmas 3. **Achenes**: Trigonous.

HABITAT AND DISTRIBUTION: Moist meadows, streamsides, and basins where snow melts late, in subalpine to alpine areas, in the higher mountains of OR and WA. AK to AB, S to CA and CO.

IDENTIFICATION TIPS: *Carex nigricans* forms a turf in snowmelt basins. Young plants with a dark, dense, oval spike have little similarity to mature plants, which have shaggy-looking inflorescences with spreading to reflexed perigynia, and with the ♀ scales fallen. Young plants are easily confused with *C. subnigricans* and cespitose *C. micropoda*, but those species differ in having narrow, involute leaves. In addition, both those species are confined to the alpine zone and have more limited ranges than *C. nigricans*. *Carex subnigricans* grows in the Wallowas and Steens Mt., and *C. micropoda* grows in the Wallowas and the mts of NW WA.

COMMENTS: *Carex nigricans* occurs in moist subalpine and alpine locations, including streamsides and open forest. It is often a community dominant in basins where deep snow melts late in the summer. Its resistance to trampling and its ability to hold soil make it useful for habitat restoration. Dense turfs of *C. nigricans* make a soft bed for backpackers. *Carex nigricans* has been used as a host plant for propagating the hemiparasite Elephant's Head (*Pedicularis groenlandica*).

top left: perigynia
*top right and center
   left and right*:
   inflorescences
bottom left: habit

Carex nudata W. Boott
Common name: Torrent Sedge
Section: *Phacocystis*   Key: G

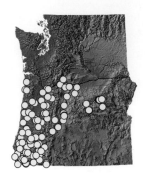

KEY FEATURES:
• Dense tussocks of gracefully arching, slender leaves
• Rocky scour zone of rivers

DESCRIPTION: Habit: Densely cespitose; the plant bases usually cannot be wrenched from their rocky strongholds. **Culms**: 35-70 cm. **Leaves**: Green or somewhat glaucous, 2-4 mm wide. Plant bases reddish brown or blackish. Leaf sheath fronts coppery, with reddish brown spots, conspicuously ladder-fibrillose. **Inflorescences**: Lowest inflorescence bract shorter than the inflorescence. Lateral 2-4 spikes ♀, 2-4.5 cm long, (4-) 5-6 mm wide, erect. Terminal 1-2 spikes ♂. ♀ **scales**: Dark reddish brown to black, much shorter than the perigynia, awnless. **Perigynia**: Pale brown with red spots and often a big black blotch near the tip, elliptical to ovate to obovate, with 3-9 veins on each face, 2.2-4 mm long, 1.2-1.8 mm wide. Beak 0.1-0.3 mm. Stigmas 2. **Achenes**: Lenticular, about half as long as the perigynia.

HABITAT AND DISTRIBUTION: Cracks in rocks or among cobbles within the scour zones of fast-moving rivers, including small rocky islands and occasionally along irrigation canals; more common in and W of the Cascades in OR, but also in some eastside river systems such as the Deschutes and John Day, in OR; also Klickitat Co., WA. S WA to N CA.

IDENTIFICATION TIPS: *Carex nudata* forms dense tussocks with narrow, gracefully drooping leaves among rocks in fast-flowing rivers. Rhizomatous *C. interrupta* may grow adjacent to *C. nudata* tussocks on stream banks in sandy or gravelly soils. *Carex nudata* in irrigation canals and road ditches can be confused with most members of section *Phacocystis,* especially if soil deposition obscures its densely cespitose habit. *Carex kelloggii* is also cespitose, but lacks ladder-fibrillose leaf sheath fronts and has green, stipitate perigynia. It grows on silts and sand but not along intensely scouring rivers.

COMMENTS: *Carex nudata* blooms early in the season, producing large amounts of seed that falls soon after spring floods. When water levels drop, the seeds are deposited in freshly scoured habitats. Seedlings can be found in early summer. Tussocks anchor rocks and capture soil that becomes a substrate for other plant species. Mature *C. nudata* tussocks usually withstand floods, and if they are dislodged, the dense, tough, bushel-sized masses of soil and roots may sprout where they land. *Carex nudata* benefits fish and the invertebrates they feed on by providing shade and hiding spots.

top left: pistillate scales, perigynia   *top right*: inflorescences
center left: ladder-fibrillose leaf sheath front   *center right*: habit
bottom: habit, habitat (rocky river scour zones)

Carex obnupta L. H. Bailey
Common name: Slough Sedge
Section: *Phacocystis*   Key: G

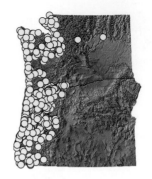

KEY FEATURES:
- Evergreen leaves
- Perigynia hard, dark, retained through winter
- Dark, arching ♀ spikes

DESCRIPTION: Habit: Rhizomatous but deep rhizomes often not collected. **Culms**: 20-120 (-200) cm. Plant bases reddish brown. **Leaves**: Evergreen, tough, green when young and aging yellowish green, 3-7 mm wide, with sharp marginal prickles pointing toward the tip. Leaf sheath fronts coppery, densely red-dotted, ladder-fibrillose. **Inflorescences**: Lowest inflorescence bract longer than the inflorescence. Lateral 2-5 spikes ♀, dark brown, erect to drooping, tapering at base, 2.5-15 cm long, 4.5-10 mm wide. Upper 2-3 (-5) spikes ♂. **♀ scales**: Reddish brown, longer than the perigynia, acute, awnless. **Perigynia**: Dark brown, hard, veinless, 2.2-3.8 mm long, 1.4-2.2 mm wide. Beak 0.1-0.3 mm. Stigmas 2. **Achenes**: Lenticular, often indented like a dented beer can on one or both sides.

HABITAT AND DISTRIBUTION: Very common in marshes, wet meadows, streamsides, pond margins, low spots in riparian woodland (often with Oregon Ash, *Fraxinus latifolia*), stabilized coastal dunes, brackish upper reaches of salt marshes, road ditches, and freshwater springs on oceanside cliffs, W of the Cascades in WA and OR, also Chelan and Okanogan cos., WA. S BC to N CA.

IDENTIFICATION TIPS: The yellowish green, arching, evergreen leaves can be identified from a moving car. In estuaries, *C. obnupta* may grow adjacent to *C. lyngbyei,* a salt marsh species with smooth-edged leaves that die in winter, yellow-brown perigynia, and straight, usually truncate-based spikes on arching peduncles. *Carex aquatilis* var. *dives* has softer leaves and perigynia, and leaf sheath fronts that lack coppery color and are not ladder-fibrillose. *Carex barbarae* of SW OR has brown, usually erect spikes and scabrous awns at the tips of the ♀ scales.

COMMENTS: *Carex obnupta* is often a community dominant. It retains its perigynia through the winter, allowing seed dispersal in spring floods. Perigynia often lack viable achenes, but we have observed seedlings in the field. *Carex obnupta* is not a preferred forage because its tough leaves have rough edges sharp enough to cut skin, but nutria and other animals can eat it. Because it is common and has long, tough rhizomes that bind soil, *C. obnupta* is often used in habitat restoration. It has potential to be invasive, especially on small sites. Split leaves have been used for basketry.

top left: pistillate scales, perigynia, achenes
top middle: immature inflorescence
top right: inflorescence
center left: ladder-fibrillose leaf sheath front
bottom left and right: habit, habitat

Carex obtusata Lilj.
Common name: Obtuse Sedge
Section: *Obtusatae*   Key: A

KEY FEATURES:
- Perigynia dark brown, contrasting with pale ♀ scales
- Single spike with spreading ♀ flowers in lower half
- Seasonally dry meadows and prairies

DESCRIPTION: Habit: Rhizomatous, the shoots arising singly or a few together on the tough, slender, dark brown rhizomes. **Culms**: (3-) 10-20 cm tall, strongly aphyllopodic, dark brown at base. **Leaves**: Flat, 0.5-1.5 mm wide, all the blades originating at the same level, at the ground surface; with minute reddish dots on the leaf sheath fronts. **Inflorescences**: One androgynous spike per culm, lacking inflorescence bracts, 0.8-1.5 cm long, the ♂ portion usually longer than the ♀ part. **♀ scales**: Brown with hyaline margins, lanceolate, 2.5-3.5 (-4.5) mm long, as wide as, and as long as or shorter than the perigynia. **Perigynia**: 1-6, ascending, tough, broadly elliptic to obovate, in cross section round or trigonous with rounded angles, grooved between the fine veins, reddish brown to dark chocolate brown, dark at maturity, 3-3.8 mm long, the beak with prominent teeth that are often white. Rachilla well developed. Stigmas 3. **Achenes**: Trigonous.

HABITAT AND DISTRIBUTION: Scree slopes, talus slopes, and ridges, above 4500 feet on the Olympic Peninsula, WA. Elsewhere, vernally moist grasslands, AK to ON, S to UT, NM, and MN, Eurasia.

IDENTIFICATION TIPS: *Carex obtusata* is a small, rhizomatous plant with dark brown perigynia contrasting with pale ♀ scales. Its dark brown rhizomes and shoot bases resemble *C. praegracilis,* but *C. praegracilis* shoot bases and rhizomes are less delicate and a blacker brown (lacking warm reddish tones). *Carex praegracilis* has pale ♀ scales nearly covering the mature perigynia and it has multiple spikes. *Carex nigricans* has broader (1.5-3 mm wide) leaves and its gradually tapered perigynia do not contrast with its dark ♀ scale. *Carex capitata* is ± cespitose and has more perigynia that are paler, have inconspicuous beak teeth, and are not filled by the achene.

COMMENTS: The *C. obtusata* plants on subalpine slopes of the Olympic Peninsula are dwarfed and very easy to overlook. Elsewhere in its range, *C. obtusata* can be a community dominant in dry prairie, particularly on silty soils and in vernally moist areas. In parts of the west, it increases when overgrazing removes taller, more palatable competitors. By default, low-yielding *C. obtusata* becomes an important forage plant in such situations.

top left: perigynia; note rachilla in dissected perigynium
top middle and right: inflorescences
center: habit
bottom: habitat (dry alpine grassland)

Carex pachycarpa Mack.
Common name: Many-rib Sedge
Section: *Ovales*   Key: J
Synonym: *Carex multicostata* Mack., misapplied

KEY FEATURES:
- Cespitose, with gynecandrous spikes and winged perigynia
- Large, dense, pale, shaggy head
- Many dorsal perigynium veins, usually impressed
- Dry mountain meadows

DESCRIPTION: Habit: Cespitose. **Culms**: 14-110 cm tall. **Leaves:** 8-25 cm long, (2) 3-5 mm wide. **Inflorescences**: 1.6-3 cm long, 12-20 mm wide, dense and head-like. Spikes gynecandrous (or rarely ♀). **♀ scales**: (3.2-) 3.9-4.8 mm, gold or light brown with paler midrib and margins, longer and narrower than the perigynia. **Perigynia**: Cream-colored or tan with green "shoulders," becoming uniformly straw-colored with age, obviously planoconvex, smoothly oval to obovate, (3.8-) 4.4-5.8 (-6.5) mm long, (1.4-) 1.8-2.8 (-3.3) mm wide, with wing (0.2-) 0.3-0.4 (-0.5) mm wide, broad near the beak but narrow lower on the body. Perigynium dorsal veins (8-) 10-17 and on dried specimens usually sunken rather than raised as in other *Ovales*, ventral veins (0-) 4-12, usually at least 3 longer than the achene. Beak winged nearly to the tip; 2-2.8 (-3.3) mm from top of achene to tip of beak. Stigmas 2. **Achenes**: Lenticular, 1.7-2.4 mm long, 1.1-1.7 mm wide, biconvex.

HABITAT AND DISTRIBUTION: Dry grasslands, mesic meadows, and open forest at moderate to high elevations in the mountains; SW OR, N in the Cascades to about Mt. Rainier, WA, and in mts E of the Cascades. WA to MT, S to CA and UT.

IDENTIFICATION TIPS: *Carex pachycarpa* is a "bunchgrass" sedge with large, pale, shaggy heads. Its pale perigynia have a recognizable but hard-to-describe shape; the main body is a smooth oval, and the wings are broad near the beak but taper smoothly to become narrow beside the achene. Many *C. pachycarpa* have impressed, rather than raised, dorsal veins. Other montane upland *Ovales* with dense heads have darker or more compact inflorescences. Many identification keys distinguish *C. pachycarpa* based only on the great number of dorsal veins on the perigynia, but occasional individuals of other species, including *C. pachystachya*, may have many veins. These other species lack the distinctive *C. pachycarpa* perigynium shape.

COMMENTS: This species has long been confused with *C. multicostata* Mack. of high elevations in the central and southern Sierra Nevada, CA. Plants of the N Warner Mts can have unusually long, lanceolate perigynia.

top left: perigynia
top right: inflorescences
center left: habit
bottom: habitat (dry montane meadow)

Carex pachystachya Cham. ex Steud.

Common name: Thick-headed Sedge
Section: *Ovales*   Key: J
Synonyms: *C. macloviana* d'Urv. var. *pachystachya* Kük.,
  C. macloviana d'Urv. ssp. *pachystachya* Hultén

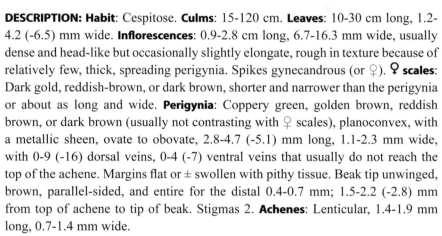

KEY FEATURES:
- Cespitose, with gynecandrous spikes and winged perigynia
- Widespread, especially at low elevations
- Compact head with rough texture due to spreading, thick perigynia
- Perigynia often coppery green

DESCRIPTION: Habit: Cespitose. **Culms**: 15-120 cm. **Leaves**: 10-30 cm long, 1.2-4.2 (-6.5) mm wide. **Inflorescences**: 0.9-2.8 cm long, 6.7-16.3 mm wide, usually dense and head-like but occasionally slightly elongate, rough in texture because of relatively few, thick, spreading perigynia. Spikes gynecandrous (or ♀). **♀ scales**: Dark gold, reddish-brown, or dark brown, shorter and narrower than the perigynia or about as long and wide. **Perigynia**: Coppery green, golden brown, reddish brown, or dark brown (usually not contrasting with ♀ scales), planoconvex, with a metallic sheen, ovate to obovate, 2.8-4.7 (-5.1) mm long, 1.1-2.3 mm wide, with 0-9 (-16) dorsal veins, 0-4 (-7) ventral veins that usually do not reach the top of the achene. Margins flat or ± swollen with pithy tissue. Beak tip unwinged, brown, parallel-sided, and entire for the distal 0.4-0.7 mm; 1.5-2.2 (-2.8) mm from top of achene to tip of beak. Stigmas 2. **Achenes**: Lenticular, 1.4-1.9 mm long, 0.7-1.4 mm wide.

HABITAT AND DISTRIBUTION: Widespread in mesic transition zones between wet and drier habitats, in moist meadows, wet prairie, marsh edges, forest edges, and roadsides; common at low elevations but extending to high elevations, throughout the PNW. Southern AK to SK, S to CA and CO, also Siberia.

IDENTIFICATION TIPS: *Carex pachystachya* has short, dense heads often with much green color. The spikes often have a star-burst look because the perigynia are thick and spreading. Similar *C. microptera* grows in slightly drier habitats and its inflorescence looks more finely textured because the perigynia are flat except where distended by the achene, so more of them are packed into each spike. See also the very similar *C. preslii, C. harfordii,* and *C. subbracteata.*

COMMENTS: *Carex pachystachya* is part of a complex of mostly self-pollinating species (e.g. *C. abrupta, C. microptera, C. preslii*), sometimes all treated as parts of *C. macloviana* d'Urv. Because they have evolved recently, probably since the last glacial retreat, they are very similar. Even within *C. pachystachya*, slightly differentiated lineages can be observed.

top left: perigynia
top right:
    inflorescences
center left: habit
center right*:
    inflorescences
bottom left: habit

Carex pallescens L.
Common name: Pale Green Sedge
Section: *Porocystis*　　　Key: F
Synonyms: *C. pallescens* L. var. *neogaea* Fern.

KEY FEATURES:
- Perigynia green and almost beakless
- Lower surfaces of leaves inconspicuously hairy
- Wet meadows

DESCRIPTION: Habit: Cespitose to short-rhizomatous. **Culms**: 20-80 cm tall, hollow, inconspicuously hairy. **Leaves**: 2-3 mm wide, hairy on the lower surface. Sheaths hairy. **Inflorescences**: Lateral spikes ♀, 0.5-2 cm long, 4.5-7 mm wide, often crowded above, the lower more distant. Terminal spike ♂, 0.5-3 cm long. ♀ **scales**: White or green to brownish, ovate, about as long as the perigynia, the lower with short, abrupt tips, the upper acute to acuminate. **Perigynia**: Green, ascending to spreading, with several faint veins, elliptic-oblong, 2.3-3 mm long, 1.1-1.5 mm wide, glabrous, beakless. Stigmas 3. **Achenes**: Trigonous, 1.8-2 mm. **Anthers**: 1.4-2.8 mm..

HABITAT AND DISTRIBUTION: Moist or mesic meadows, ditches, and forest openings; probably native in NE WA and N ID (where found at ~2300 feet); introduced to Clark Co., SW WA, and extreme SW BC. MN to NF, S to WV and MA, disjunct in the PNW. Also native to Europe; introduced to New Zealand.

IDENTIFICATION TIPS: This inconspicuous species can be mistaken for a grass. Similar *C. hirsutella*, recently introduced to Linn Co., OR, has gynecandrous terminal spikes, longer anthers, and more densely hairy foliage. Both have beakless perigynia, unlike our other sedges with hairy leaves. *Carex sheldonii* and *C. hirta* are rhizomatous and have densely hairy leaves. *Carex whitneyi* is cespitose and has longer perigynia.

COMMENTS: The idea that the NE WA and N ID populations of *C. pallescens* are native is plausible because they grow in relatively undisturbed wet meadows with other apparently native plants that also occur in eastern N America. Introduced populations are to be looked for in disturbed wet meadows elsewhere in the PNW. *Carex pallescens* is an early successional species that declines as shrubs and trees produce a closed canopy. However, it has a long lasting seed bank and reappears after disturbance. It is a host of arbuscular mycorrhizal fungi. A cultivar with white-edged leaves has been developed for ornamental use.

top left: pistillate scales, perigynia
top right: inflorescences
bottom: habit

Carex pansa L. H. Bailey

Common name: Sand-dune Sedge
Section: *Divisae*   Key: B, H
Synonyms: *C. arenicola* Fr. Schmidt ssp. *pansa* (L. H.
Bailey) T. Koyama & Calder

KEY FEATURES:
- Coastal sand dune habitats
- Separate ♂ and ♀ plants
- Blackish rhizomes and shoot bases

DESCRIPTION: Habit: Rhizomatous, the rhizomes blackish, stout, 1.8-2.6 mm in diameter, with shoots arising singly or in small clusters. **Culms**: 8-40 cm. Shoot bases dark brown to blackish. **Leaves**: 1.7-3.8 mm wide, mainly clustered at the base of the plant. **Inflorescences**: 1.2-2.5 cm long, of many short spikes. Plants apparently dioecious; flowers in one plant seemingly all ♀ or all ♂ but usually with a few, nearly undetectable flowers of the opposite sex mixed in the spikes. ♀ **scales**: Dark reddish brown to purplish black with hyaline margins, similar in color to or darker than the perigynia, awned or not, hiding the perigynia. **Perigynia**: Dark brown to nearly black, ovate to obovate, 3.1-4.2 mm long, 1.3-1.8 mm wide. Beak 0.7-1.5 mm long. Stigmas 2. **Achenes**: Lenticular. **Anthers**: With a tiny bristly awn.

HABITAT AND DISTRIBUTION: Coastal sand dunes and sandy meadows, BC to CA.

IDENTIFICATION TIPS: *Carex pansa* forms a green or yellow-green turf on stabilized sand dunes and in the lawns of beachfront houses. Its little-branched, blackish rhizomes produce straight lines of evenly spaced shoots. Its ♀ inflorescences are large and dark. *Carex praegracilis,* widespread E of the Cascades and in SW OR, is extremely similar, but has lighter colored and narrower ♀ inflorescences. Another sedge of coastal sand dunes is *C. brevicaulis,* which has much-branched, brown rhizomes, reddish shoot bases, and pubescent perigynia.

COMMENTS: *Carex pansa* helps stabilize coastal sand dunes. It may form a dense turf or weave sparsely through other vegetation such as Kinnikinnick (*Arctostaphylos uva-ursi*). It has been planted as a substitute for lawn grasses on well-drained soils, though it may become aggressive. *Carex pansa* lawns tolerate trampling, need little mowing, and thrive on less water than most lawn grasses, though they may need some watering to prevent summer dormancy. Some "*C. pansa*" lawn cultivars are misidentified *C. praegracilis.* Only confirmed *C. pansa* should be used near the coast to prevent gene flow from *C. praegracilis* into *C. pansa.*

Carex pansa

top left: pistillate scales, perigynia
top right: pistillate inflorescence
center left: habit
bottom: habitat (stabilized coastal sand)

Carex pauciflora Lightf.
Common name: Few-flower Sedge
Section: *Leucoglochin*   Key: A

KEY FEATURES:
- Single spike
- Perigynia long, narrow, spreading to reflexed
- Plants extremely delicate

DESCRIPTION: Habit: Long-rhizomatous, the culms arising singly or a few together. **Culms**: 10-40 (-60) cm long, delicate, longer than or as long as the leaves. **Leaves**: The lower ones bladeless, those with blades only 1-3 per culm, 0.5-1.6 mm wide. **Inflorescences**: One spike per culm, (0.3-) 0.5-0.8 (-1) cm long, androgynous, lacking inflorescence bracts. ♀ **scales**: Deciduous, about 5 mm long and shorter than the perigynia. **Perigynia**: Light green to yellowish brown, ascending when young but soon spreading to reflexed, lanceolate, finely single-nerved, (5-) 5.9-7.8 mm long, 0.7-1.1 mm wide, with persistent and exserted styles. Stigmas 3. **Achenes**: Trigonous.

HABITAT AND DISTRIBUTION: *Sphagnum* bogs, generally growing in moss mats; Puget Sound and North Cascades of WA. AK to NB, S to NW WA, WV, and CT.

IDENTIFICATION TIPS: *Carex pauciflora*'s needle-like, spreading perigynia give it a unique appearance. *Carex anthoxanthea* and *C. circinata* also have very narrow perigynia, but the perigynia ascend and the ♀ scales are persistent. *Carex circinata* is cespitose in rocky habitats and has involute leaves, while *C. anthoxanthea* has much shorter perigynia.

COMMENTS: This delicate species can be difficult to see, growing mixed with other vegetation. Nonetheless, it is a community dominant in some bog communities in AK. *Carex pauciflora* has scattered populations across the southern portion of its range, remnants of a wider distribution during the ice ages. It is rare in the PNW and some populations seem to be declining. Its seed dispersal is interesting. When fully ripe, the perigynia point downward due to the growth of spongy tissue on the ventral side of the perigynium base. Any touch by a passing animal may lift the tip of the ripe perigynium and compress the spongy base. When pressure is released, the perigynium springs up, flying up to 2 feet from its origin.

top left, top right, center right, bottom left: inflorescences, perigynia
bottom right: habit, habitat
(*Sphagnum* bog)

Carex paysonis Clokey

Common name: Payson's Sedge
Section: *Scitae*   Key: F
Synonyms: *C. podocarpa* R. Brown, misapplied, *C.
    podocarpa* R. Brown var. *paysonis* (Clokey)
    B. Boivin, *C. tolmiei*, misapplied

KEY FEATURES:
• Perigynia broad, with marginal ribs displaced from
    the margins
• Terminal spike ♂
• Moist subalpine to alpine habitat

DESCRIPTION: Habit: ± rhizomatous. **Culms**: 15-50 cm tall, usually phyllopodic.
Leaves: 3-6 mm wide. **Inflorescences**: 3-5 spikes, each 0.5-2.5 cm. Lateral spikes
♀, the upper ones erect and the lower ones drooping, often on long peduncles.
Terminal spike ♂ (occasionally androgynous). ♀ **scales**: Light brown to reddish
black, margins dark or rarely hyaline, midvein dark (occasionally light), narrower
and longer or shorter than the perigynium, acute or with a short pointed tip, but not
truly awned. **Perigynia**: Ovate or obovate to round, purple-black at least toward
the tip, flattened, veined, with the marginal ribs displaced away from the margin,
3.5-4 mm long, usually more than half as wide as long, smooth, with beak 0.3-0.5
mm long. Stigmas 3. **Achenes**: Trigonous, filling half or less of the perigynium
bodies.

HABITAT AND DISTRIBUTION: Subalpine and alpine meadows, talus, and snowmelt
basins, the sites generally moist or if apparently dry then with underground
moisture, above 6800 feet; NE OR; reported in error from WA. S BC to SK, S to
OR, UT, and WY.

IDENTIFICATION TIPS: *Carex paysonis* is a rhizomatous, montane sedge, with a
staminate terminal spike and broad perigynia with prominent ribs located near
but not on the margins. Similar *C. spectabilis* has elliptic perigynia with the 2
marginal ribs usually (but not always) truly marginal, and has aphyllopodic fertile
shoots. Intermediates between *C. spectabilis* and *C. paysonis* exist in NW MT,
and the 2 species can be difficult to distinguish wherever their ranges overlap.
Most of the other species in section *Racemosae* (e.g. *C. heteroneura*) differ from
C. paysonis in having a gynecandrous terminal spike.

COMMENTS: *Carex paysonis* is used for high elevation habitat restoration and mine
reclamation projects because it can be a community dominant, holds soil well,
and tolerates heavy-metal contamination. Establishmment is greater with plugs
rather than seeds, but sometimes it is planted on such high-acid, low-nutrient mine
tailings that even this tough plant cannot survive.

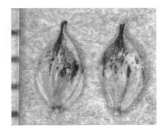

top left and top middle:
    perigynia
*top right and bottom
    right*: inflorescences
center left: habit

Carex pellita Muhl. ex Willd.

Common name: Woolly Sedge
Section: *Paludosae*   Key: C
Synonyms: *C. lanuginosa* Michx., misapplied

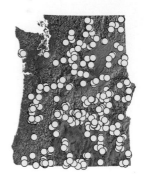

KEY FEATURES:
- Perigynia pubescent
- Leaf sheaths reddish, varnished-looking, glabrous, fibrillose
- Long-rhizomatous, in wetlands

DESCRIPTION: Habit: Rhizomatous. **Culms**: 15-70 cm tall. **Leaves**: Yellow-green, flat or M-shaped in cross section, (2-) 2.2-4.5 (-6) mm wide, the tips not curled. Leaf sheaths glabrous, rarely very short-pubescent, lower ones shiny reddish, ladder-fibrillose. **Inflorescences**: Inflorescence bract erect or nearly so, longer than the inflorescence. Lateral 1-3 spikes ♀, erect or ascending, well separated from the 1-3 terminal ♂ spikes. ♀ **scales**: Glabrous or with scabrous margin apically, acute to awned. **Perigynia**: Ascending, broadly ovoid, densely pubescent, 2.4-5.2 mm long, 1.7-2.8 mm wide, abruptly contracted to the beaks which are (0.6-) 0.8-1.6 mm long, with straight teeth 0.4-0.8 mm long. Stigmas 3. **Achenes**: Trigonous.

HABITAT AND DISTRIBUTION: Marshes, stream banks, ditches, pond margins, and wet meadows, often in places that are dry at the surface in summer, often in habitats with a history of disturbance, at low to mid-montane elevations; widespread in WA and OR. AK to NF, S to CA, TX, and VA, Mexico.

IDENTIFICATION TIPS: *Carex pellita* is a slender wetland plant, sometimes forming dense stands but often mixing inconspicuously with other vegetation. Its perigynia are densely pubescent and taper abruptly to short beaks. Its slender, erect inflorescent bracts surpass the staminate spike. Sterile plants can be identified by the glabrous, shiny, varnished-looking red-brown shoot bases with ladder-fibrillose leaf sheath fronts. Very similar *C. lasiocarpa* is a taller plant that grows mainly in bogs and fens. It occupies deeper water when the two grow together. Its leaves are narrower, rarely as much as 3 mm wide, channeled or involute near the base and almost solidly triangular in cross section near the tip. Its lowest inflorescence bracts diverge at about 45 degrees from the culm. *Carex sheldonii* has longer perigynia with longer teeth and has ± pubescent leaf sheath fronts and leaves.

COMMENTS: *Carex pellita* is planted in wetland restoration projects, using either plugs or seeds. Its seeds are eaten by diverse birds. Native Americans used its rhizomes for basketry. Historically, *C. pellita* was called *C. lanuginosa,* but the type specimen for that name was eventually identified as an immature *C. lasiocarpa,* making the name unavailable for this species.

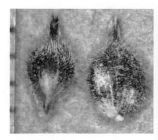

top left: perigynia
top middle: spike
top right: inflorescence
center left: habit
bottom left: ladder-fibrillose leaf
    sheath front

Carex pelocarpa F. J. Herm.

Common name: Duskyseed Sedge
Section: *Racemosae*   Key: F
Synonyms: *C. nova* L. H. Bailey var. *pelocarpa* (F. J.
 Hermann) Dorn

KEY FEATURES:
• Large, blackish, compact, shiny heads on slender
 culms
• Rocky, mesic, alpine slopes

DESCRIPTION: Habit: Cespitose. **Culms**: 10-40 cm tall. **Leaves**: 3-4 mm wide. **Inflorescences**: 2-4 sessile spikes, all crowded together or rarely the lowest one remote. Lateral spikes ♀, all of them short, 0.7-1 cm long, 5-7 mm wide. Terminal spike gynecandrous. **♀ scales**: Dark brown or black, including the margins and midvein, lanceolate, about as long as the perigynia, with apex acute or mucronate. **Perigynia**: Glossy dark reddish brown, often with lighter, yellowish or green margins, obovate or circular, nearly flat, 3.5-4.5 mm long, 2.5-3 mm wide, abruptly narrowed to a beak 0.5-0.8 mm long. Stigmas 3. **Achenes**: Trigonous, filling less than half of the perigynium body.

HABITAT AND DISTRIBUTION: Alpine slopes, ridges, boulder fields, talus slopes, and on rocky streamsides, in damp but well-drained soils, usually on sites with no competing vegetation. Wallowas and Steens Mt., OR. OR to MT, CO, and NV.

IDENTIFICATION TIPS: *Carex pelocarpa* is an alpine sedge with dark, sessile spikes. It is similar to *C. media,* which also has sessile spikes, but *C. media* grows in lower elevation, often wetter habitats and has smaller, green perigynia that are nearly filled by the achene. Its spikes usually have more contrast of dark scales with light perigynia, in part because the scales are usually less than half as long as the perigynia. *Carex albonigra* has the lower spikes on short, easily overlooked stalks, and has white ♀ scale margins.

COMMENTS: *Carex pelocarpa* has been considered a synonym of *C. nova*, which has a similar dark, dense inflorescence. *Carex nova* has narrower perigynia that differ in being distallly papillose and having serrulate beak margins. Its ♀ scales are about as wide as its perigynia. It grows in wetter montane to alpine meadows and streambanks. *Carex nova* is a more southern species not found in the PNW, but the two taxa grow in the same mountain ranges in parts of the Great Basin.

Carex pelocarpa

top left: pistillate scales, perigynia

top right and center left: inflorescences

bottom: habit, habitat (alpine fellfield)

Carex pendula Huds.
Common name: Pendulous Sedge
Section: *Rhynchocystis*   Key: F

KEY FEATURES:
• Huge, clump-forming sedge
• Very long, narrow, drooping spikes
• Moist, shaded habitats

DESCRIPTION: Habit: Cespitose, forming large clumps that may be a foot or more across at the base. **Culms**: 100-200 cm long, often spreading at an angle rather than erect. **Leaves**: 80-130 cm long, 8-20+ mm wide. **Inflorescences**: Lateral spikes ♀, drooping, 10-30+ cm long, 5-8 mm wide. Terminal spike ♂, arching. ♀ **scales**: Brown with pale midrib, as long as or a little longer than the perigynia. **Perigynia**: 2.6-4 mm long, green to brown, elliptic, with a beak 0.5 mm long. Stigmas 3. **Achenes**: Trigonous.

HABITAT AND DISTRIBUTION: Shaded streamsides and ravines in riparian forest, capable of growing in almost any wet soil but thriving in high clay soils; escaped and naturalized in the Seattle and the Willamette Valley and likely to spread widely in riparian forests. Native to Europe, NW Africa, and SW Asia; introduced to VA, IL, WA, OR, CA and probably other states, also New Zealand.

IDENTIFICATION TIPS: *Carex pendula* is a giant, clump-forming sedge with long, drooping spikes. Sterile shoots resemble those of *Scirpus microcarpus* and *C. amplifolia,* but those species are rhizomatous. *Carex amplifolia* differs in having paler leaves, erect culms with ascending spikes and perigynia with longer beaks.

COMMENTS: In Europe, *C. pendula* is a common, often dominant, species of moist woodland habitats. It is sold in the PNW for use in the garden, and variegated cultivars have been developed. Although mature plants are truly impressive, *C. pendula* should not be planted in the PNW because it sets seed abundantly and invades nearby shaded riparian zones. Landowners have found it nearly impossible to eradicate from yards. It has been observed spreading in sidewalk cracks and along shaded creeks in urban parks and nearby rural areas. *Carex pendula* has the potential to become a serious pest in PNW riparian forests.

top left: perigynia
top middle: spike
top right: inflorescence
center left: habit
center right: inflorescence
bottom left: habit, habitat (riparian)

Carex petasata Dewey
Common name: Liddon Sedge
Section: *Ovales*   Key: J

KEY FEATURES:
• Cespitose, with gynecandrous spikes and winged
  perigynia
• Large, strongly veined perigynia
• Sagebrush steppe and other dry sites

DESCRIPTION: Habit: Cespitose. **Culms**: 30-85 cm. **Leaves**: 10-30 (-40) cm long, 2-4 (-5) mm wide. **Inflorescences**: (2-) 2.5-4.5 (-6) cm long, 9-16 mm wide, erect, somewhat elongate, the spikes distinct. Spikes gynecandrous. ♀ **scales**: 5.8-7.6 mm long, dark gold or red-brown with white margins 0.2-0.7 mm wide, acute, approximately as long and wide as the perigynia and concealing the bodies and more than half the beak. **Perigynia**: Appressed to ascending, whitish green to brown, lanceolate to ovate, 6-8 mm long, 1.7-2.4 mm wide, with 10+ dorsal veins and 4-10 ventral veins. Beak tip unwinged, brown, parallel-sided, and entire for up to 1 mm; or sometimes winged and ciliate-serrulate almost to the tip; (2.8-) 3-4.6 mm from top of achene to beak tip, dorsal suture white-edged and relatively conspicuous. Stigmas 2. **Achenes**: Lenticular, 2.2-3 mm long, (1.1-) 1.3-1.8 mm wide, 0.5-0.7 mm thick.

HABITAT AND DISTRIBUTION: Sagebrush steppe, dry to moist meadows, and open forest from about 2500 feet to timberline; E of the Cascades in WA and OR. AK to NWT, S to CA, AZ, and CO.

IDENTIFICATION TIPS: *Carex petasata* is a large species with ascending spikes in an erect inflorescence, and large perigynia that are mostly concealed by the ♀ scales. It may be confused with *C. praticola,* which has smaller perigynia and usually a nodding inflorescence. *Carex xerantica* has smaller perigynia and white (to yellowish) ♀ scales, and grows in the northern Great Plains. All reports of *C. xerantica* from the PNW are *C. petasata* or *C. praticola.* See account of the rare *C. davyi,* which has perigynia as long as those of *C. petasata.* Its ♀ scales are shorter, revealing most of the beak, and have no white margins or very narrow ones.

COMMENTS: *Carex petasata* cover of about 2-3% is considered an indicator of ranges in excellent condition. The plants are palatable to livestock, especially early in the season when foliage is still succulent. Whether or not individual plants survive grazing, the early season grazing reduces or prevents seed set. Failure to reproduce eventually causes the population to decline. *Carex petasata* is absent or rare in large areas of potentially suitable habitat.

top left: perigynia
top, center, and bottom right: inflorescences
bottom left: habit

Carex phaeocephala Piper

Common name: Dunhead Sedge, Alpine Hare Sedge
Section: *Ovales*   Key: J

KEY FEATURES:
- Cespitose, with gynecandrous spikes and winged perigynia
- Spikes overlapping, pointing stiffly upward in a narrow inflorescence
- Perigynia translucent, pale or brown, veinless or faintly veined
- Dry, alpine slopes and grasslands

DESCRIPTION: Habit: Cespitose. **Culms**: (5-) 15-45 cm tall. **Leaves**: 5-20 cm long, 1-2.5 (-3) mm wide, folded, somewhat tough, the edges sometimes revolute. **Inflorescences**: 1.5-3.5 (-4) cm long, 10-15 mm wide, somewhat elongated although all the spikes overlap, erect or somewhat angled to one side. Spikes gynecandrous, narrow, ascending to erect. ♀ **scales**: Gold to brown with paler midrib, 3.7-5.1 mm long, ± covering the perigynia. **Perigynia**: Gold to (occasionally) dark brown, translucent, ovate, 3.5-4.5 (-5.2) mm long, 1.5-2.3 mm wide, veinless (or with up to 5 faint ventral veins and up to 9 dorsal veins). Beak tip unwinged, brown, and parallel-sided for the distal 0.6 mm, with the perigynium wings extending up to the base of the cylindric tip, giving the perigynium a somewhat boat-like shape; 1.8-2.2 mm from beak tip to achene top. Stigmas 2. **Achenes**: Lenticular, 1.5-2 mm long, 0.8-1.1 (-1.2) mm wide, 0.3-0.5 mm thick.

HABITAT AND DISTRIBUTION: Dry, rocky alpine or subalpine slopes in OR and WA. AK to AB, S to CA and CO.

IDENTIFICATION TIPS: *Carex phaeocephala* is a tough, cespitose sedge of rocky alpine slopes, with overlapping, ascending spikes, and ♀ scales that hide the perigynia. Its somewhat translucent, ovate perigynia are usually veinless or have faint veins on the ventral surface. In general appearance, it is identical to *C. tahoensis,* which grows at slightly lower elevations and has larger perigynia that are tough, strongly veined, opaque, and brown. Most keys and descriptions are suspect because the two species have been confused. *Carex praticola* perigynia are similar in texture, color, and number of veins, but that species has longer perigynium beaks and nodding inflorescences, and grows in moist meadows at lower elevations.

COMMENTS: As is true for other alpine sedges, *C. phaeocephala* seedlings grow rapidly, a necessity for establishment during a short growing season. The plant is used for revegetation of disturbed alpine sites.

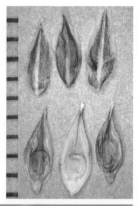

top left: pistillate scales, perigynia
top middle and top right: inflorescences
center left: habit
bottom right: habitat (dry alpine ridge)

Carex pluriflora Hultén
Common name: Black Bog Sedge
Section: *Limosae*   Key: F

KEY FEATURES:
- Black scales nearly hiding pale to blackish perigynia
- Lower spikes dangling on long peduncles
- Bog habitat

DESCRIPTION: Habit: Rhizomatous, the shoots arising singly or in small groups. Young roots with yellow, dense, felty hairs. **Culms**: 20-60 cm tall. **Leaves**: 1.5-4 mm wide, glaucous, flat, margins sometimes revolute. **Inflorescences**: Lowest inflorescence bract 0.5-2.5 mm long, bristle-like. Lateral spikes ♀, 1.2-2 cm long, 6-8 mm wide. Terminal spike ♂, 1-2.5 cm long, 2.2-3 mm wide, rarely gynecandrous. **♀ scales**: Black, 3.5-4.5 mm long, 2.1-3.8 mm wide, slightly wider and shorter than the perigynia, with apex acute to acuminate, mucronate or with a short awn to 1.5 mm long. **Perigynia**: Pale green to dark brown or blackish at maturity, papillose, 3.2-4.2 mm long, 1.5-2.6 mm wide, beakless. Stigmas 3. **Achenes**: Trigonous.

HABITAT AND DISTRIBUTION: Bogs, marshes, fens, wet meadows, streamsides, and lakeshores, 0-3200 feet elevation; NW WA, and one coastal bog in NW OR. AK to OR, and NE Asia.

IDENTIFICATION TIPS: *Carex pluriflora* is a spiky, rhizomatous sedge with few, ascending leaves and dangling, black ♀ spikes. Similar *C. limosa* has brown ♀ scales. *Carex magellanica* has narrow ♀ scales that are awned and longer than the perigynia.

COMMENTS: *Carex pluriflora* can be a dominant species in bogs with *Sphagnum* moss and thick organic soils, often on seasonally flooded sites where competition from shrubs is limited. Although it grows in freshwater sites, it is most likely to be a community dominant in coastal bogs that receive some flooding by tidewater, and it may grow into the *C. lyngbyei* zone. It is tolerant of disturbances such as seasonal flooding, light grazing, and fire, and requires occasional disturbance to prevent succession to shrubs. In WA, threats include pollution, trampling by livestock intensive enough to alter its habitat, and heavy recreational use. At the one OR site, succession to a coastal shrub community is displacing *C. pluriflora*.

Carex pluriflora

top left: perigynia
top middle: pistillate scales, perigynia
top right: inflorescence
center left: felty yellow root hairs
bottom left: habit

Carex praeceptorum Mack.
Common name: Teachers' Sedge
Section: *Glareosae*   Key: I
Synonyms: *C. praeceptorium* Mackenzie

KEY FEATURES:
- Small, almost beakless, brown perigynia
- Perigynium with dark brown dorsal suture, darker than the rest of the beak
- High elevation bogs and streamsides

DESCRIPTION: Habit: Loosely or densely cespitose. **Culms**: 10-20 (-30 cm long. **Leaves**: 1.5-2.5 mm wide, green or gray-green, shorter than the culms. **Inflorescences**: Usually brownish, (1-) 1.5-2 cm long with 3-5 spikes, all spikes crowded or sometimes the lowest one somewhat remote, gynecandrous, oblong, 0.4-0.7 cm long with 8-15 (-20) perigynia. ♀ **scales**: Brown with whitish centers and margins, about as long as the perigynia. **Perigynia**: Ascending, light brown (sometimes dark brown with age), with several distinct veins, (1.9-) 2.0-2.3 (-2.4) mm long, elliptic, with short beak. Beaks blunt, to 0.5 mm long. Dorsal suture about as long as the beak, dark brown, darker than the surrounding perigynium surface. Stigmas 2. **Achenes**: Lenticular.

HABITAT AND DISTRIBUTION: Bogs, wet meadows, wet edges of lakes and streams, in mountains, 7200-9000 feet in OR, above 5800 feet in WA, E of crest of Cascades in WA and in mts of E OR. BC to MT, S to CA and CO.

IDENTIFICATION TIPS: *Carex praeceptorum* is a small, cespitose plant with brownish inflorescences with overlapping spikes. Its dorsal suture is about as long as the beak, dark brown, and darker than the rest of the perigynium. *Carex canescens* is usually quite different with gray-green leaves and perigynia. However, some variants of *C. canescens* have green foliage and brown perigynia; these can easily be misidentified as *C. praeceptorum*. They differ in having the lower spikes separated and the dorsal suture about the same color as the rest of the beak.

COMMENTS: *Carex praeceptorum* is a cute, inconspicuous sedge of high elevation bogs and streamsides. It seems to be uncommon or rare. The major threats to its existence are hydrologic changes to its habitat. Preeminent American caricologist Kenneth K. Mackenzie named this sedge *C. praeceptorum* (which translates as "sedge of teachers") to honor Oregon professors Morton Peck and James C. Nelson, who sent him many specimens.

top left: perigynia
top middle: habit
top right: inflorescence
bottom left: habitat (high-elevation wet
    meadow/bog)
bottom right: habit

Carex praegracilis W. Boott
Common name: Blackcreeper Sedge
Section: *Divisae*   Key: B, H

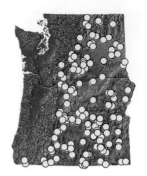

KEY FEATURES:
- Separate ♂ and ♀ plants
- Blackish rhizomes and shoot bases
- Seasonally dry, alkaline or serpentine areas

DESCRIPTION: Habit: Rhizomatous, the rhizomes blackish, stout, 1.8-3 mm in diameter, with shoots arising singly or in small clusters. **Culms**: (10-) 25-80 cm. Shoot bases dark brown to blackish. **Leaves**: 1-3 (-3.5) mm wide, mainly clustered at the base of the plant. **Inflorescences**: 0.9-4 cm long with short spikes. Plants apparently dioecious; flowers in one plant seemingly all ♀ or all ♂ but usually with a few, nearly undetectable flowers of the opposite sex mixed in the spikes. ♀ **scales**: Pale, straw-colored to pale reddish brown with hyaline margins, more or less hiding the perigynia, similar in color to or paler than the perigynia, awned or not. **Perigynia**: Variable in color and shape, brown to nearly black, sometimes paler, ovate to obovate, (2.2-) 2.6-3.7 mm long, 1.2-1.9 mm wide, the base often filled with pithy tissue. Beak 0.7-1.2 mm long. Stigmas 2. **Achenes**: Lenticular. **Anthers:** With a tiny bristly awn.

HABITAT AND DISTRIBUTION: Usually found in alkaline or serpentine soils, in seasonally moist prairies and plains. Elevation usually low to moderate, but may grow up to 10,000 feet; E of the Cascades in OR and WA, also SW OR. Common from YT to MB, S to N CA, Mexico, and KS, and spreading along roads that are salted in winter E to QC and VA. Introduced to central and S CA.

IDENTIFICATION TIPS: *Carex praegracilis* is strongly rhizomatous and ± dioecious. Its nondescript pistillate inflorescences hide the perigynia. Rhizomes and shoot bases are blackish; shoots arise singly. *Carex douglasii,* which grows with *C. praegracilis,* has slender brown rhizomes and broader, ovoid, ♀ inflorescences. It tolerates more alkaline conditions. *Carex simulata* grows in truly wet places and has stubby, short-beaked perigynia. Near the coast, see *C. pansa.*

COMMENTS: Where common, *C. praegracilis* is an important winter or early spring forage for cattle, horses, and wildlife, because this tough little plant survives grazing, trampling, and poor soils better than many more palatable and higher-yielding species. On some alkaline ranges it has been harvested for hay. It is planted as an "eco-friendly" lawn because it tolerates trampling, requires little mowing, and thrives on little water, but it can spread aggressively. It should not be planted near the coast; see *C. pansa.*

Carex praegracilis

top left: perigynia
top right: pistillate inflorescences
center left: rhizome and dark stem
  bases
center right: staminate
  inflorescences
bottom left: habit, habitat (moist
  alkaline meadow)

Carex praticola Rydb.

Common name: Meadow Sedge
Section: *Ovales*   Key: J
Synonyms: *C. piperi* Mackenzie

KEY FEATURES:
- Cespitose, with gynecandrous spikes and winged perigynia
- Inflorescence elongated, nodding
- Large perigynia
- Moist to mesic meadows

DESCRIPTION: Habit: Cespitose. **Culms**: (16-) 35-70 (-100) cm. **Leaves**: 10-30 cm long, (1.5-) 2-3 (-4) mm wide. **Inflorescences**: (1.7-) 2.5-5 cm long, 10-15 (-17) mm wide, bent to the side, somewhat elongate, the spikes distinct and the lowest one remote. Spikes gynecandrous. ♀ **scales**: Straw-colored, reddish brown, or brown with similar or paler midstripe, (3.4-) 4.2-5.8 mm long, as long as or longer than but narrower than the perigynium body, with white margin 0.1-0.3 mm wide. **Perigynia**: Appressed to ascending, with margins brown distally but often whitish over the achene, (3.7-) 4.5-6 mm long, 1.2-2 mm wide, usually with 4-11 dorsal veins and 0-4 ventral veins, and with wings (0.1-) 0.2-0.4 (-0.5) mm wide. Beak relatively long, the tip unwinged, brown, parallel-sided, and entire for the distal 0.4-1 mm, with the very tip white; (1.6-) 1.9-3 mm from achene top to beak tip. Stigmas 2. **Achenes**: Lenticular, 1.4-2.1 (-2.7) mm long, (0.8-) 1-1.5 mm wide, 0.4-0.6 mm thick

HABITAT AND DISTRIBUTION: Moist to mesic meadows, and forest openings, often in somewhat disturbed spots, at low to moderate elevations; widespread in mts of OR and WA and scattered W of the Cascades. AK to NF, S to CA, CO, and ME, native in N ON. Introduced in S ON and IL.

IDENTIFICATION TIPS: *Carex praticola* has a greenish to pale brown, nodding or bent inflorescence with white-margined ♀ scales. *Carex petasata* has longer perigynia. W of the Cascades, similar *C. leporina* has ♀ scales that are not white-margined, though they may have white-hyaline tips or bases. Its perigynium beaks are brown at the very tip. Its perigynia average shorter (3.4-5.2 mm long) with shorter beaks (1.5-2 mm from achene top to beak tip), and are usually veined on the ventral surface.

COMMENTS: *Carex praticola* is likely under-reported. It seems to be much more widespread than previously recognized. Plants W of the Cascades may be recent introductions because the first Willamette Valley, OR, record is from 1916. Alternatively, the species may be a Willamette Valley native that has undergone a severe decline since settlement.

top left: pistillate
   scales, perigynia,
   achenes
top middle: perigynia
top right and
   center right:
   inflorescences
bottom: habit,
   habitat

Carex preslii Steud.
Common name: Presl's Sedge
Section: *Ovales*   Key: J

KEY FEATURES:
- Cespitose, with gynecandrous spikes and winged perigynia
- Like a wimpy *C. pachystachya*
- Dry, montane uplands

DESCRIPTION: Habit: Cespitose. **Culms**: 23-55 cm. **Leaves**: 6-30 cm long, 1.7-3.6 mm wide. **Inflorescences**: 1.4-3 cm long, (5-) 8-15 mm wide, head-like but somewhat elongated because the lower spikes are slightly removed from the upper ones (lowest internode 3-7 mm). Spikes gynecandrous (or ♀). **♀ scales**: 2.8-3.7 mm, gold to red-brown, occasionally white, with a green, off-white, or dark midrib, shorter or longer and narrower than the perigynia, with white margin 0.1-0.2 (-0.5) mm wide. **Perigynia**: Green, straw-colored, or golden-brown with green margins that contrast in color with the ♀ scales, lacking metallic sheen, ovate to obovate, (3.3) 3.5-4.3 mm long, (1.3-) 1.5-2 mm wide, with 0-5 ventral veins. Beak winged nearly to the tip, or unwinged, brown, parallel-sided, and entire for up to 0.6 mm (both types mixed in a single inflorescence), 1-2.4 mm from top of achene to tip of beak. Stigmas 2. **Achenes**: Lenticular, 1.5-2 mm long, 1-1.5 mm wide, 0.4-0.7 mm thick..

HABITAT AND DISTRIBUTION: Roadsides, dry open slopes, forest edges, and mesic subalpine meadows, at moderate to high elevations in mts, Cascades, Olympic Mts., NE WA, NE and SW OR. AK S to CA and ID.

IDENTIFICATION TIPS: Squeeze all the perigynia out of a suspected *C. preslii* head or two to see the diversity of its perigynium beaks. That diversity and a drier habitat are the main features distinguishing *C. preslii* from a runty *C. pachystachya*. All *C. pachystachya* perigynia have the beak tips unwinged, brown, and parallel-sided for 0.4-0.7 mm. The latter species tends to have more and larger spikes in a more compact inflorescence, its darker perigynia (golden brown or coppery to dark brown) have a metallic sheen that *C. preslii* lacks, and it grows in a more mesic meadow or meadow edge habitat.

COMMENTS: Even the most devoted sedgeheads roll their eyes at this species, which is part of a complex of recently evolved, mostly self-fertilizing taxa, including *C. pachystachya*. *Carex preslii* has diverged in its habitat requirements and presumably in its physiology but with little morphological change. It is really hard to distinguish from related species. Its tendency to grow on roadsides suggests it may be useful for road decommissioning and other habitat restoration projects.

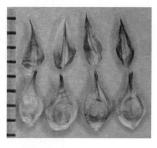

top left: pistillate scales, perigynia
top right: inflorescences
center left and right: habit
bottom: habitat

Carex proposita Mack.
Common name: Potato Chip Sedge
Section: *Ovales*   Key: J

KEY FEATURES:
- Cespitose, with gynecandrous spikes and winged perigynia
- Wide, flat perigynia with broad, crinkled margins
- Plants short, with folded leaves
- Dry alpine ridges

DESCRIPTION: Habit: Cespitose. **Culms**: 10-35 cm. **Leaves**: 6-20 cm long, (0.5-) 1-2 (-2.5) mm wide if flat, but often channeled or folded. **Inflorescences**: 1.7-3 cm long, 11-18 mm wide, reddish brown, the 3-6 spikes distinguishable but usually crowded, the lowest internode generally 4-8.6 mm. Spikes gynecandrous. ♀ **scales**: Light brown with pale midstripe, 3.5-4.8 mm long, shorter than and a third to half as wide as the perigynia, the tip obtuse to acuminate. **Perigynia**: Light brown or reddish brown, the margins broadly green or whitish above and sometimes at the base, flat except over the achene, broadly ovate, widest at or above the middle, abruptly narrowed to the small beak, 4.3-6.3 mm long, 2-3 mm wide, with 3-11 dorsal veins and 0-13 ventral veins, with wings 0.4-0.9 mm wide, the edge often crinkled. Beak up to 1 mm long, unwinged, brown, parallel-sided, and entire for the distal (0.3-) 0.5-0.9 mm; (1.9-) 2.2-3.7 mm from achene top to beak tip. Stigmas 2. **Achenes**: Lenticular, located in the center of the perigynium, 1.5-2.1 mm long, 0.7-1.1 (-1.3) mm wide, 0.2-0.5 mm thick. **Anthers:** Long-persistent.

HABITAT AND DISTRIBUTION: Among rocks on dry alpine ridges, (4500-) 6200-8000 ft.; Wenatchee Mts. and Cascades, WA. Scattered in ID, WY, and CA, reported from BC.

IDENTIFICATION TIPS: This is a tough, short, high elevation sedge with broad, flat, "potato chip" perigynia. The other high elevation member of section *Ovales* with similar perigynia is *C. straminiformis,* but its perigynia are a different shape, widest below the middle (of total perigynium length) and with the achene low in the perigynium body. It also has broader, flat leaves and its inflorescence is straw-colored and blackish, not reddish. *Carex haydeniana* and *C. microptera* have longer perigynium beaks, inflorescences that are dark, dense heads with spikes not easily distinguishable, and broader leaves.

COMMENTS: This alpine sedge has a disjunct range and is rare in the PNW. Some populations are threatened by recreational use of their alpine habitat.

top left: perigynia
top right: inflorescence
center left: perigynia
bottom left and right:
    habit

Carex raynoldsii Dewey
Common name: Raynolds' Sedge
Section: *Racemosae*   Key: F

KEY FEATURES:
- Plump, orangish perigynia
- Moist sites in mountain meadows

DESCRIPTION: Habit: Cespitose, often loosely so. **Culms**: 20-75 cm tall. **Leaves**: 3-7 mm wide. **Inflorescences**: 3-6 erect to ascending spikes, these usually overlapping and clustered near the tip of the culm, but sometimes the lowest remote. Lateral spikes on short stalks (not sessile), ♀, 1-2 cm long, 3-5 mm wide. Terminal spike ♂. **♀ scales**: Dark brown or blackish, the margin and sometimes the midvein lighter, shorter than or as long as the perigynia, narrower than the perigynia at least above, mostly acute (but varying from obtuse to acuminate). **Perigynia**: Greenish yellow to yellow-brown, sometimes gray, often orange-brown above, contrasting with the darker ♀ scales, elliptic or obovate, (3-) 3.3-4.5 mm long, 1.75-2 mm wide, unusually thick and swollen-looking, with an abrupt beak 0.3-0.5 mm long. Stigmas 3. **Achenes**: Trigonous, 1.8-2.4 mm long, nearly filling the perigynium bodies, yellowish brown when mature.

HABITAT AND DISTRIBUTION: Moist to dry meadows, often associated with seeps, in mts to near timberline; in and E of the Cascades in WA and OR, also on the Olympic Peninsula, WA. S BC to AB, S to CA and CO.

IDENTIFICATION TIPS: *Carex raynoldsii* is a cespitose sedge with short spikes containing plump, often orangish perigynia that contrast with the dark scales. *Carex spectabilis* and *C. paysonis* may grow with *C. raynoldsii,* but have flatter perigynia that are nearly hidden by the ♀ scales and are not filled by the small achenes. *Carex raynoldsii* could be confused with *C. aboriginum*, which has perigynia (4.7-) 5-6 mm long and is very rare, found only in W ID.

COMMENTS: The plump achenes make *C. raynoldsii* an acceptable trailside nibble for humans. This species is sometimes common enough in upland habitats to be an important forage for elk, cattle, horses, sheep, and, where present, bison. Like most cespitose sedges and grasses, it decreases under heavy grazing. *Carex raynoldsii* could make a valuable contribution to ecosystem restoration projects in appropriate habitats.

top left: perigynia

top right: inflorescence

center left: Inflorescences with smut

center right: habit

center left: inflorescences

bottom: habitat (montane meadow, here with American Bistort, *Bistorta bistortoides*)

Carex retrorsa Schwein.
Common name: Retrorse Sedge
Section: *Vesicariae*   Key: E

KEY FEATURES:
- Spikes crowded
- Inflorescence bracts much longer than inflorescence
- Big, inflated perigynia
- Partially shaded wetlands

DESCRIPTION: Habit: Densely cespitose. **Culms**: 10-100 cm tall. **Leaves**: green, flat to W-shaped, the widest leaves 3-10 mm wide. Basal leaf sheaths often brown or tinged reddish. **Inflorescences**: Inflorescence bracts 3-9 times longer than inflorescence. Lateral spikes ascending to spreading, ♀ or androgynous, (1.5-)2-4.6 cm, crowded or with the lowest somewhat separated. Terminal 1 (-3) spike(s) ♂, not or barely elevated above the lateral spikes. ♀ **scales**: Shorter than the perigynia, apex acute to acuminate, awnless. **Perigynia**: Spreading or mostly reflexed when mature, green or straw-colored, with 6-13 veins, 6-10 mm long, (1.6-) 2.1-3.4 mm wide, the apex contracted to a distinct beak 2.1-4.5 mm long. Stigmas 3, style persistent, becoming curved. **Achenes**: Trigonous.

HABITAT AND DISTRIBUTION: Floodplain forests, edges between lakes and forests, swamps, streamsides, wet thickets, wet meadows, in sites that are soggy wet in spring but may dry out on the surface later in the season, scattered locations including Kittitas Co. and NE WA; Eugene, the Columbia River downstream of Hood River, and NE OR. BC to NB, S to OR, UT, and PA, not present in the Great Plains.

IDENTIFICATION TIPS: *Carex retrorsa* is a robust, cespitose sedge with its spikes all nestled down near the bases of its long inflorescence bracts. Its ♂ spike is located among the clustered ♀ spikes. One's first impression might be that this is a deformed *C. utriculata* or *C. vesicaria* with an abnormally short inflorescence. Both of these species have spikes more separated, with the ♂ spikes held distinctly above the ♀ ones, and with inflorescence bracts not more than 2.5 times as long as the inflorescence. *Carex utriculata* is strongly rhizomatous, forming large stands.

COMMENTS: Most PNW populations are isolated and very small. E of the Great Plains, *Carex retrorsa* is planted in rainwater control structures that are not subject to much siltation. It is also grown as an attractive garden subject.

top left and center left: perigynia; note
 contorted persistent style in dissected
 perigynium
top right: inflorescence
bottom: habit

Carex rossii Boott
Common name: Ross' Sedge
Section: *Acrocystis*   Key: C

KEY FEATURES:
- Pubescent perigynia
- Basal ♀ spikes
- Perigynium beak usually 1+ mm long

DESCRIPTION: Habit: Cespitose, often loosely so, or short-rhizomatous, the shoots usually arising in small clusters. **Culms**: 7-30 (-40) cm tall. Plant bases reddish, with ladder-fibrillose sheaths. **Leaves**: Green, rarely glaucous in SW OR, but not densely papillose on the lower surface. **Inflorescences**: Bract of lowest lateral spike leaf-like, longer than the inflorescence. Lateral spikes (just below the staminate spike) 1-4, ♀, 0.5-0.9 (-1.1) cm long, with 3-10 perigynia. Terminal spike ♂, (0.5-) 0.6-1.3 cm long. There are also 1-2 basal ♀ spikes among the leaf sheaths. ♀ **scales**: Pale to dark reddish brown with white margins, shorter than the perigynia, apex acute to acuminate. **Perigynia**: Pubescent, elliptic to obovoid, with succulent bases that wither when dry, veinless except for the two ribs, perigynia from lateral (not basal) spikes 3.1-4.5 mm long, 1.4-1.7 mm wide, with beaks 0.9-1.7 mm long, with apical teeth 0.2-0.4 mm long. Stigmas 3. **Achenes**: Trigonous.

HABITAT AND DISTRIBUTION: Uplands, including open forests, sagebrush steppe, meadows, grassy coastal headlands, roadsides, cut banks, and clearcuts; throughout WA and OR. AK to ON and MI, S to CA and NM.

IDENTIFICATION TIPS: *Carex rossii* is common, forming green, leafy clumps in upland habitats. It has very short basal ♀ spikes nestled among the shoot bases. Remnants of the basal spikes can be found long after its early spring flowering. Other species with basal spikes are: *C. brainerdii* (glaucous, leaves papillose SW OR), *C. brevicaulis* (with leathery leaves, strictly coastal), and *C. deflexa* var. *boottii* (short perigynium beaks, mainly above 3000 feet). *Carex inops* ssp. *inops,* common in and W of the Cascades, lacks short basal ♀ spikes, has short inflorescence bracts, and is more strongly rhizomatous.

COMMENTS: The basal spikes of *C. rossii* are a hedge against herbivory and present the perigynia to ants, which disperse them. *Carex rossii* provides forage for pocket gophers, bighorn sheep, domestic sheep, and bears. Cattle graze it selectively in spring. It is an early-successional species, thriving after fire or clearcutting. It is resistant to trampling and its extensive root system holds soil well. Variation in leaves, habit, and beaks suggest that plants we now call *C. rossii* include two species.

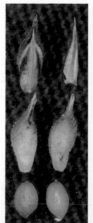

top left: pistillate scales, perigynia, achenes
top right: inflorescence
center left and bottom left: habit

Carex rostrata Stokes
Common name: Northern Beaked Sedge
Section: *Vesicariae*   Key: E

KEY FEATURES:
- Leaves strongly glaucous and papillose
- Leaf sheath with crosswalls between veins, giving appearance of brickwork
- Bog habitat in NE WA

DESCRIPTION: Habit: Rhizomatous, forming large stands. **Culms**: 8-90 cm tall **Leaves**: Glaucous, U-shaped in cross section with involute margins, strongly papillose (at 20X) on the upper surface, 1.5-4.5 (-7.5) mm wide. Basal leaf sheaths often brown to tinged reddish, somewhat spongy, with crosswalls common between the longitudinal veins. **Inflorescences**: (1-) 2-3 erect to ascending ♀ spikes, mostly 2-10 cm long, with 8 spiraling columns of perigynia. Terminal (1-) 2-4 spikes ♂, 2-7 cm long, well separated from the ♀ spikes. ♀ **scales**: Mostly shorter than the perigynia, apex acute to acuminate, usually awnless. **Perigynia**: Spreading, green, straw-colored, or reddish brown, with 9-15 veins, 3.6-5.8 mm long, 1.7-2.8 mm wide, abruptly narrowed to a beak (1-) 1.2-2 mm long, with straight teeth 0.2-0.7 mm long. Stigmas 3, style persistent, becoming curved. **Achenes**: Trigonous.

HABITAT AND DISTRIBUTION: Fens and bogs, NE WA. AK to NF, S to WA, MT, MN, and ME; Eurasia.

IDENTIFICATION TIPS: *Carex rostrata* forms dense stands of relatively tall, grayish plants. Its corncob-like spikes and rhizomatous habit are very similar to widespread *C. utriculata,* which has leaves that are broader, green, and smooth or very sparsely papillose. Both *C. rostrata* and *C. utriculata* have many crosswalls between the vertical veins on the backs and sides of the leaf sheaths, giving the sheaths a pattern like brickwork.

COMMENTS: *Carex rostrata* can be a community dominant, and it is an important forage for cattle in northern areas, e.g., Iceland and Siberia. In the PNW, populations are very few and small, remnants left behind as glaciers retreated northward. This species and *C. utriculata* have often been considered one species because their growth habit and inflorescences are nearly identical. The differences in leaf surface and anatomy are so pronounced and consistent that they are now considered different species. Where the two species grow together in NE WA, the differences in leaf color and width are conspicuous. Most older floras identify *C. utriculata* as *C. rostrata,* thus most of the literature on ecology of N American "*C. rostrata*" is actually about *C. utriculata.* Before basing management on research reports, one should clarify which species was studied.

top left: perigynia and pistillate scales
top right: inflorescences
center left: sheaths with brick pattern
bottom: habit (rhizomatous, bluish
    plants), habitat (bog)

Carex saxatilis L.

Common name: Russet Sedge
Section: *Vesicariae*   Key: E, F, G
Synonyms: *C. physocarpa* C. Presl, *C. saxatilis* var. *major* Olney

KEY FEATURES:
- Perigynia glabrous, with 2 stigmas and persistent styles
- Rhizomatous with reddish brown shoot bases
- High elevation wetlands with peaty soils

DESCRIPTION: Habit: Rhizomatous. **Culms**: 8-90 cm tall. **Leaves**: Green, flat to V-shaped, 0.9-5 mm wide. Basal leaf sheaths reddish brown. **Inflorescences**: Lateral 1-3 ♀ spikes, 1-3 cm long, erect or ascending or the lower one often nodding on its slender peduncle. Terminal 1-3 spikes ♂, 1-4 cm long, well separated from the ♀ spikes. **♀ scales**: As long as or shorter than the perigynia, apex acute to acuminate, awnless. **Perigynia**: Shiny, ascending, green to straw-colored or often reddish brown to blackish, with 2 ribs and otherwise veinless or with obscure dorsal veins, 2.2-5.5 mm long, 1.1-2.9 mm wide, with the apex abruptly narrowed to a short beak 0.2-0.8 mm long, with straight teeth 0.3 mm long. Stigmas 2 (-3), style persistent, usually becoming curved. **Achenes**: Lenticular (or trigonous).

HABITAT AND DISTRIBUTION: Fens, bogs, lakeshores, at high elevations; N Cascades and Olympics of WA, NE OR, Lake Co., OR. AK to NF, S to OR, CO, and ME.

IDENTIFICATION TIPS: *Carex saxatilis* is a slender, rhizomatous sedge, forming solid stands or mingling inconspicuously with other wetland plants. The narrow terminal spike is well separated from the ♀ spike(s). These plants are most likely to be interpreted at first as an odd-looking something else. Their narrow shoots and reddish brown shoot bases recall *C. pellita* and *C. lasiocarpa,* which have densely pubescent perigynia. Shiny perigynia and persistent styles differentiate *C. saxatilis* from *C. aquatilis* and its allies in section *Phacocystis.*

COMMENTS: *Carex saxatilis* is rare in the PNW, where it is at the southern edge of its range. Major threats are climate change and recreational use of habitat. To the N, this species is used for erosion control and for hay. Historically, N American plants have been divided into two or more species, but the E-W variation is continuous, so they are best treated as one taxon. *Carex saxatilis* hybridizes with *C. vesicaria* and *C. utriculata.* The hybrids are usually sterile and morphologically intermediate between their parents. Some of their perigynia have 2 stigmas and some have 3. Because they spread by rhizomes, the hybrids can persist and form large populations although they rarely set seed.

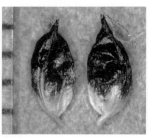

top left: achene with contorted persistent style
top middle: perigynia
top right: inflorescences
center left: shoot bases
botttom: habit

Carex scabriuscula Mack.

Common name: Siskiyou Sedge
Section: *Scirpinae*   Key: A, (B, C)
Synonyms: *C. gigas* (T. Holm) Mackenzie

KEY FEATURES:
- Usually dioecious
- Serpentine substrates in SW OR and N CA
- Mature spike dark purplish brown; usually 1 per stem

DESCRIPTION: Habit: Loosely cespitose, with short rhizomes. **Culms**: 30-65 cm tall, about twice as tall as leaves, phyllopodic. **Leaves**: 2-3 mm wide; bases persist into following year. **Inflorescences**: Plants usually dioecious; 1 (occasionally 2 or 3) spike per stem; 20-40 perigynia per spike. ♀ **scales**: Reddish brown to purple, lanceolate, shorter than and about half as wide as the perigynia. **Perigynia**: 2.5-4 mm long, 1-2.5 mm wide, broadly elliptic; purplish to blackish at maturity; slightly hairy on distal third; beak 0.5 mm long. Stigmas 3 (occasionally 4). **Achenes:** Trigonous (or quadrangular). Mature achene up to 2.8 mm long, 1.4 mm wide, filling only from about a third to a little less than two thirds the length and width of the perigynium.

HABITAT AND DISTRIBUTION: Moist to mesic serpentine substrates in meadows, lake margins, and seeps in open forest, 900-6000 feet elevation; Curry and Josephine cos., OR and NW CA, disjunct in Plumas Co., CA.

IDENTIFICATION TIPS: *Carex scabriuscula* forms dense mounds of leaves in moist, serpentine habitats. Few similar sedges grow in these habitats in SW OR. *Carex mendocinensis* has a terminal ♂ spike and lateral ♀ spikes, the lowest of them well down the stem, not just terminal or near-terminal as in *C. scabriuscula*. *Carex scirpoidea* ssp. *pseudoscirpoidea* is dioecious like *C. scabriuscula*, but is shorter, does not have persistent leaf bases, and has narrower perigynia that are nearly filled by mature achenes.

COMMENTS: *Carex scabriuscula* is rare and was considered endemic to the Klamath Ecoregion until plants were found on the W slope of the Sierra Nevada in Plumas Co., CA. It is vulnerable to habitat changes resulting from succession, as conifers encroach on its moist meadow habitat. Harmful effects are compounded where people camp at lakeshore conifer groves, trampling the plants.

The type specimen was collected by William Cusick in 1902 in a "wet meadow of the Cascade Mountains," a term which in those days could include the Siskiyous. Although the collection number fits a Cascades trip he made between Ashland and the Klamath Basin, there is no serpentine in that area, and searches for it there have been unsuccessful.

top left: pistillate
   scales, perigynia
top right: pistillate
   inflorescence
center left: pistillate
   and staminate
   inflorescences
bottom: habit, habitat

Carex scirpoidea Michx. ssp. *pseudoscirpoidea* (Rydb.) D. A. Dunlop

Common name: Western Singlespike Sedge,
Section: *Scirpinae*   Key: A, (B, C)
Synonyms: *C. pseudoscirpoidea* Rydberg

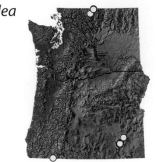

KEY FEATURES:
- Usually separate ♂ and ♀ plants
- Usually a single spike per culm
- Perigynia ovate and hairy toward the tip

DESCRIPTION: Habit: Rhizomatous. **Culms**: 5-31 cm long, erect, surrounded by the previous year's dead leaves at the base. **Leaves**: To 3.5 mm wide (narrower in SW OR). **Inflorescences**: Dioecious, 1 (occasionally 2 or 3) spike per stem, cylindrical. ♀ spikes 1.5-4 cm long. ♀ **scales**: Reddish brown to purple, ovate, 2.8 mm long, usually shorter than or as long as the perigynia, with hyaline margin that may be fringed with hairs, sometimes hairy on the back. **Perigynia**: Ovate, short-hairy especially toward the beak, (1.5-) 2-2.8 (-3) mm long, 1.5 mm wide, tightly enveloping the achenes. Stigmas usually 3 (occasionally 2). **Achenes**: Usually trigonous (occasionally lenticular), 1.5-1.8 mm long, 0.9-1.2 mm wide, filling the perigynia.

HABITAT AND DISTRIBUTION: Mesic to dry ridges and fellfields, generally in areas that have subsoil moisture, often due to snowmelt; often among rocks; on a variety of substrates including calcareous soils and rarely serpentine, usually subalpine to alpine, generally above 6000 feet, Okanogan Co., WA, Steens Mt. and S Jackson Co. in OR. BC to MT, S to CA and CO.

IDENTIFICATION TIPS: *Carex scirpoidea* forms leafy, more or less spreading clumps or turf and produces cylindrical, unisexual usually solitary spikes. The perigynia and often ♀ scales are covered with hairs. *Carex scirpoidea* spp. *pseudoscirpoidea* is distinguished from the other subspecies by its smallish perigynia and the remnants of the previous year's leaves at the bases of the flowering culms. In SW OR and NW CA, *C. s.* ssp. *pseudoscirpoidea* can be confused with *C. scabriuscula,* which is confined to serpentine substrates and is a taller plant with larger blackish perigynia. Immature plants are hard to distinguish.

COMMENTS: In *C. scirpoidea* ssp. *pseudoscirpoidea,* the fertile culm begins growth the year prior to flowering, and dead leaves of the previous year can be found at its base. In the other *C. scirpoidea* subspecies, the flowering culm completes its growth during the same season that it flowers. Therefore, all leaves on the flowering culm are typically alive at flowering.

Carex scirpoidea ssp. *pseudoscirpoidea*

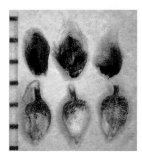

top left: pistillate scales, perigynia
top right: pistillate inflorescences
center right: staminate inflorescences
bottom left: habit
bottom right: habitat

Carex scirpoidea Michx. ssp. *scirpoidea*
Common name: Northern Singlespike Sedge
Section: *Scirpinae*   Key: A, (B, C)

KEY FEATURES:
- Usually separate ♂ and ♀ plants
- Usually a single spike per culm
- Perigynia ovate, and hairy toward the tip

DESCRIPTION: Habit: Cespitose, with short rhizomes. **Culms**: (5-) 10-35 (-40) cm tall, erect. Flowering culms aphyllopodic, without remains of previous year's leaf blades at the base, with lowest leaves reduced to bladeless sheaths. **Leaves**: 2.5 mm wide. **Inflorescences**: Dioecious, 1 (occasionally 2 or 3) spike per stem, cylindrical. ♀ **scales**: Reddish brown to purple, ovate, 2.5 mm long, usually shorter than the perigynia, with hyaline margin that may be fringed with hairs, sometimes hairy on the back also. **Perigynia**: Ovate, short-hairy especially toward the beak, (1.8-) 2-2.5 (-3) mm long, 1-1.2 (-1.5) mm wide, tightly enveloping the achenes. Stigmas 3 (occasionally 2). **Achenes**: Trigonous (occasionally lenticular), 1.5-1.8 mm long, 0.8-1.2 mm wide, filling the perigynia.

HABITAT AND DISTRIBUTION: Mesic to dry meadows, ridges and fellfields, generally in areas that have subsoil moisture due to snowmelt; usually subalpine to alpine, usually on calcareous soils but also on granodiorite and serpentine; N WA, NE OR. AK to NF, S to OR, CO, and NY; Norway, E. Russia.

IDENTIFICATION TIPS: *Carex scirpoidea* produces cylindrical, usually solitary, unisexual spikes. The perigynia and often ♀ scales are covered with hairs. The subspecies *scirpoidea* is distinguished from the other *C. scirpoidea* subspecies by its smallish perigynia (mostly 2-3 mm long and 1.5-2 times as long as wide) and the bladeless sheaths at the base of its flowering shoots. Rare *C. idahoa,* known from Grant Co., OR, may have superficially similar spikes but its perigynia and ♀ scales are glabrous and plants are strongly rhizomatous, forming a turf.

COMMENTS: The three subspecies of *C. scirpoidea* have a continuum of ecological requirements. *Carex s.* ssp. *pseudoscirpoidea:* highest, mostly alpine elevations, driest sites, on diverse substrates. *Carex s.* ssp. *scirpoidea:* alpine to montane, intermediate moisture, on limestone or other non-acidic substrates. *Carex s.* ssp. *stenochlaena:* low to moderate elevation, wet, weakly acidic sites, not limestone. The subspecies interbreed in the greenhouse but are mostly distinct in the field.

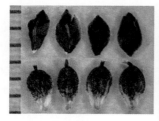

top left: pistillate scales, perigynia
top right: pistillate inflorescences
bottom right: habit (staminate plant
   left, pistillate plant right

Carex scirpoidea Michx. ssp. *stenochlaena*
(T. Holm) Á. Löve & D. Löve

Common name: Alaska Singlespike Sedge
Section: *Scirpinae*   Key: A, (B, C)
Synonyms: *C. stenochlaena* (T. Holm) Mackenzie

KEY FEATURES:
- Usually separate ♂ and ♀ plants
- Usually a single spike per culm
- Perigynia lanceolate, and hairy toward the tip

DESCRIPTION: Habit: Cespitose, with short rhizomes. **Culms**: 24-34 cm long, lax, those of ♀ inflorescences, in particular, tending to droop. Flowering culms aphyllopodic, without remains of previous year's leaf blades at the base, with lowest leaves reduced to bladeless sheaths. **Leaves**: 2.5 mm wide. **Inflorescences**: Dioecious, 1 (occasionally 2 or 3) spike per stem, somewhat shaggy-looking, widest near the top, loosely flowered below. ♀ **scales**: Reddish brown to purple, lanceolate, 3.5 mm long, often longer than the perigynia, with hyaline margin that may be fringed with hairs, sometimes hairy on the back also. **Perigynia**: Lanceolate, short-hairy especially toward the beak, dark purplish, (2.8-) 3-4 (-5) mm long, 0.9-1.4 (-1.6) mm wide, tightly enveloping only the lower ¾ of the achene. Stigmas usually 3 (occasionally 2). **Achenes**: Usually trigonous (occasionally lenticular), 1.2-2 mm long, 0.8-1.2 mm wide, not filling the distal portion of the perigynium.

HABITAT AND DISTRIBUTION: Moist meadows, streambanks, and rocky slopes, the more southerly populations on waterfalls and seepy cliffs, on somewhat acidic substrates, from low elevations to about 4000 feet, N Cascades and Olympics, WA, and W Cascades in Lane Co., OR; AK to OR, also MT.

IDENTIFICATION TIPS: *Carex scirpoidea* ssp. *stenochlaena* forms loose, leafy tufts with cylindrical, unisexual spikes that may be solitary or with 1 or 2 smaller spikes at the base of the main spike. The perigynia and often ♀ scales are covered with hairs. The subspecies *stenoclaena* can be distinguished by its longer, narrower perigynia and its preferences for wet, somewhat acidic habitats. Some plants from WA are intermediate between this subspecies and *C. s.* ssp. *scirpoidea.*

COMMENTS: Rare in OR, this is the most common of the three *C. scirpoidea* subspecies in WA. Threats to populations include changes to the hydrology of its habitats. One OR population is on a seepy roadside cliff and could be threatened by road construction or roadside herbicide spraying. All three subspecies are uncommon to rare in the PNW but most populations are probably secure, often occurring in wilderness areas.

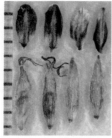

top left: pistillate scales, perigynia
top right: inflorescences (4 pistillate, 2 staminate)
center left: perigynium
center right: habit
bottom: habit, habitat

Carex scoparia Schkuhr ex Willd. var. *scoparia*
Common name: Pointed Broom Sedge
Section: *Ovales*   Key: J

KEY FEATURES:
- Cespitose, with gynecandrous spikes and winged perigynia
- Spikes very fine textured, with acuminate ♀ scales and narrow, flat perigynia
- Inflorescence angled to the side or arching
- Marshes, wet meadows, mostly W of the Cascades

DESCRIPTION: Habit: Cespitose. **Culms**: 20-100 cm, the vegetative culms occasionally producing shoots and roots at the nodes and functioning as stolons. **Leaves**: 10-32 cm long, 1.4-3 (-3.5) mm wide. Leaf sheath front narrowly white-hyaline, often hidden in dried specimens. **Inflorescences**: 1.5-6 cm long, 5-20 mm wide, erect to angled or arching, spikes distinguishable although often overlapping. Spikes gynecandrous. ♀ **scales**: Hyaline brown, often with a green midstripe, 3.4-4 mm long, shorter and much narrower than the perigynia, the tip acuminate. **Perigynia**: Ascending, green to light brown, lanceolate and at least 3 times as long as wide, flat except over the achene, 4.2-5.5 (-6) mm long, 1.2-2 mm wide, 0.35-0.5 mm thick; wings 0.2-0.6 mm wide, with the widest points on each side usually at different distances along the perigynium length, and much narrower below the middle; with about 5 veins on each face, or fewer on the ventral side. Beak tip winged and ciliate-serrulate almost to the tip; 2.2-4.8 mm from achene top to beak tip. Stigmas 2. **Achenes**: Lenticular, 1.3-1.7 mm long, 0.7-0.9 mm wide, 0.3-0.4 mm thick.

HABITAT AND DISTRIBUTION: Marshes and wet meadows, generally growing in water or saturated soils; in WA and OR mainly W of the Cascades, N from Douglas Co., OR, also NE WA. BC to SK, S to CA and MT, also MB to NF, S to GA and CO. Introduced to Europe, Australia, and New Zealand.

IDENTIFICATION TIPS: *Carex scoparia* is a greenish (to straw-colored) sedge with very fine-textured spikes due to the relatively flat, appressed perigynia and the narrowly pointed tips of both the ♀ scales and the perigynium beaks. It is frequently confused with *C. leporina*, which has coarser-textured spikes and smaller perigynia, and *C. feta,* a more erect plant with coarser-textured spikes.

COMMENTS: *Carex scoparia* seed is collected in the northern Willamette Valley, propagated, and used in wetland restoration projects, rainwater control ponds, and wildlife plantings, at a rate of 1-10% of the total seed mix. Most commercially available seed should not be planted in the PNW because it originates in the E and Midwest. Seeds are eaten by diverse birds.

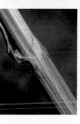

top left: perigynia
top right: inflorescence
center left: habitat (on
   cracking mud)
center right: leaf sheath
   front
bottom: inflorescences

Carex scopulorum Holm var. *bracteosa* (L. H. Bailey) J. F. Herm.

Common name: Mountain Sedge
Section: *Phacocystis*   Key: G

KEY FEATURES:
- Blackish ♀ scales and perigynia
- Short, clustered spikes
- Glaucous leaves
- Montane wetlands

DESCRIPTION: Habit: Rhizomatous, but rhizomes short and plants sometimes clump-forming. **Culms**: 11-65 cm. **Leaves**: Glaucous, 3-6 mm wide. Plant bases reddish brown to purplish black. Lower leaf sheaths to about 5 cm long, with blades. Leaf sheath fronts hyaline with light brown or purple spots, not ladder-fibrillose. **Inflorescences**: Lowest inflorescence bract usually shorter than the inflorescence. Lateral 2-4 spikes ♀, erect, usually not tapered at base, 1-2.5 cm long, 3-5 mm wide. Terminal 1-2 spikes ♂. **♀ scales**: Dark purplish or blackish, about as along as the perigynia. **Perigynia**: Obovate, green or brown with reddish brown spots, usually extensively blotched with purplish black on the upper half, veinless, 2-4 mm long, 1.2-2.3 mm wide, the tip obtuse to rounded. Beak 0.2-0.3 mm long, not bidentate. Stigmas 2 (3). **Achenes**: Lenticular (trigonous).

HABITAT AND DISTRIBUTION: Seasonally wet habitats including streamsides and wet meadows, high montane to alpine (more common below timberline), in partial shade to full sun; Cascades of WA and OR and mts of E and SW OR. YT to AB, S to CA and CO.

IDENTIFICATION TIPS: Despite great variability in perigynium color and spike length, *C. scopulorum* is recognizable because of its montane habitat, glaucous leaves, and blackish inflorescence. *Carex spectabilis,* in dryish to mesic sites near and above timberline, is similar but always has 3 stigmas/perigynium and spikes often less crowded, the lower sometimes dangling. Vegetatively, *C. scopulorum* resembles lower-elevation *C. aquatilis,* which usually has broader leaves, more separated spikes, and the inflorescence bract longer than the inflorescence. In NW WA, compare with *C. scopulorum* var. *prionophylla.*

COMMENTS: *Carex scopulorum* is a common, community dominant sedge of wet habitats in mountains, suitable for habitat restoration in montane wetlands. It decreases under heavy grazing. Its inflorescences often bulge with the blue-gray fruiting bodies of the smut fungus *Anthracoidea bigelowii.* Short young shoots produce haloes of conspicuous yellow anthers when freed by snowmelt, sometimes as late as August.

Carex scopulorum var. bracteosa

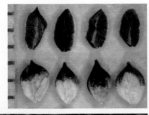

top left: pistillate scales, perigynia
top right: inflorescences
center left: immature inflorescence
bottom: habit, habitat (montane, moist
    meadow)

Carex scopulorum T. Holm var. *prionophylla* (T. Holm) L. A. Standley
Common name: Firethread Sedge
Section: *Phacocystis*   Key: G

KEY FEATURES:
- Blackish ♀ scales and perigynia
- Short spikes
- Long, bladeless lower leaf sheaths

DESCRIPTION: Habit: Rhizomatous and sometimes sod-forming. **Culms**: 35-90 cm. **Leaves**: Glaucous, 3-6 mm wide. Plant bases reddish brown to purplish black. Lower leaf sheaths to about 15 cm long, lacking blades. Leaf sheath fronts reddish brown with red or pale brown spots, ladder-fibrillose. **Inflorescences**: Lowest inflorescence bract usually shorter than the inflorescence. Lateral 2-4 spikes ♀, erect, usually not tapered at base, 1-2.5 cm long, 3-5 mm wide. Terminal 1-2 spikes ♂. **♀ scales**: Dark purplish brown, about as long as the perigynia. **Perigynia**: Elliptic to obovoid, brown with reddish brown spots, usually extensively blotched with purplish black on the upper half, veinless, 2-4 mm long, 1.2-1.7 mm wide, the tip acute. Beak 0.2-0.3 mm long, not bidentate. Stigmas 2. **Achenes**: Lenticular.

HABITAT AND DISTRIBUTION: Wet habitats well below timberline in the mountains, N-central and NE WA. BC to MT, ID, and WA.

IDENTIFICATION TIPS: *Carex scopulorum* var. *prionophylla* is distinguished from the more widespread and common *C. scopulorum* var. *bracteosa* because the latter has shorter lower leaf sheaths that have normal blades, veinless leaf sheath fronts, and perigynia that are obtuse to rounded apically. Populations intermediate between the two varieties occur in NE OR. Even in extreme NE WA, many plants that should probably be identified as variety *prionophylla* may have leaf sheaths not quite as long as they "should" be, and perigynia of variable shape.

COMMENTS: *Carex scopulorum* var. *prionophylla* can be a densely rhizomatous community dominant in montane wetlands in a limited area that includes N-central and NE WA. It has potential for use in erosion control and habitat restoration projects, but its range is limited.

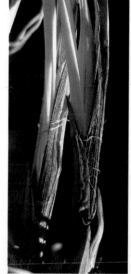

top left: pistillate scales, perigynia
top right: inflorescences
center left: aphyllopodic shoots with ladder-fibrillose
    leaf sheaths
bottom: habit, habitat (montane, moist meadow)

Carex serpenticola Zika
Common name: Serpentine Sedge
Section: *Acrocystis*   Key: A, C

KEY FEATURES:
- Pubescent perigynia
- Shoots often either all ♂ or all ♀
- Serpentine substrates in SW OR, NW CA

DESCRIPTION: Habit: Rhizomatous, sometimes forming mats or rings. **Culms**: 8-40 cm tall, short and erect at anthesis but the female flowering culms elongated in fruit and arching to the ground. Plant bases purplish black. **Leaves**: Green, usually longer than the culms, (1.5-) 2.2-3.5 (-5) mm wide, not papillose on the lower surface. **Inflorescences**: Lowest inflorescence bract usually bristle-like, usually shorter than the inflorescence. Plants often dioecious but sometimes with a ♂ (or gynecandrous) terminal spike and ♀ lateral spikes. Solitary ♂ spikes 1.1-2.4 cm long. ♂ inflorescences often with a small, infertile ♀ spike at the base. Lateral spikes ♀, 0-3, 0.6-1.1 cm long. Terminal ♀ spikes 1-1.5 cm long. Basal ♀ spikes usually none, or if present on elongated stalks. **♀ scales**: Purplish black, white-margined, with paler midrib, longer and narrower than the perigynium. **Perigynia**: Pubescent, green, ripening tan or dark purple, obovate, with succulent bases that wither when dry, veinless except for the 2 marginal ribs (rarely with up to 5 veins to mid-length), 3.1-3.6 mm long, 1.5-1.8 mm wide; beaks 0.5-1.0 mm long, with apical teeth 0.2 mm long. Stigmas 3. **Achenes**: Globose-trigonous.

HABITAT AND DISTRIBUTION: Serpentine savanna, in seasonally moist uplands, road ditches, abandoned roadbeds, dry creek terraces, and edges of seeps; Curry and Josephine cos., OR, and Del Norte Co., CA.

IDENTIFICATION TIPS: *Carex serpenticola* is a short, upland species with pubescent perigynia and purplish black leaf sheaths and ♂ and ♀ scales. It flowers in April or earliest May, and perigynia fall off by June. No similar species is normally dioecious. *Carex rossii* and *C. brainerdii* have basal ♀ spikes nestled among the reddish shoot bases. *Carex serpenticola* may grow in mixed populations with similar *C. concinnoides,* which has 4 stigmas and thus perigynia that are four-angled at least near the base, reddish brown shoot bases, and a more open growth form with more arching leaves.

COMMENTS: Because *C. serpenticola* blooms so early, its dioecious habit went unnoticed by botanists. It was not described until 1998. It is a narrow endemic. Some populations are threatened by mining, but overall it is secure at this time. Ants drag the perigynia away by their succulent bases, dispersing the seeds.

top left: pistillate scales, perigynia   *top right*: inflorescences
center left: inflorescences   *center right*: habit
bottom: habitat (serpentine savanna)

Carex serratodens W. Boott
Common name: Two-tooth Sedge
Section: *Racemosae*   Key: F

KEY FEATURES:
- Spikes short and thick
- Terminal spike usually gynecandrous
- Moist, often partly shaded meadows on ± serpentine substrates

DESCRIPTION: Habit: Cespitose. **Culms**: 30-120 cm long, often sprawling. **Leaves**: 2-4 mm wide. **Inflorescences**: 3-5 spikes, each 0.8-3 cm long, the lateral spikes ♀, the terminal spike usually gynecandrous (sometimes ♂). ♀ **scales**: Light (or dark) brown, hyaline-margined, with midvein and broad center area conspicuous, lighter in color than the body, the midvein sometimes raised and/or with minute spines, apex acute to mucronate, sometimes with prominent hairs. **Perigynia**: Broadly elliptical or ovate, green or light brown, veined, (2-) 3-5 mm long, 1.2-2 mm wide, with beak 0.5-1 mm long, with two small teeth. Stigmas 3. **Achenes**: Trigonous, nearly filling the body of the perigynia.

HABITAT AND DISTRIBUTION: Moist meadows, hillsides, and seeps, in sun or more often in partial shade, often on serpentine substrates, at low to moderate elevations, Douglas Co., OR, S to CA and AZ.

IDENTIFICATION TIPS: *Carex serratodens* can be recognized by its cespitose habit, short, thick spikes, gynecandrous terminal spikes, striped ♀ scales, and toothed perigynia. *Carex gynodynama* is somewhat similar but has much wider, hairy leaves and longer, narrower, usually darker spikes. Its terminal spike is usually ♂. *Carex mendocinensis* has longer, narrow spikes; the terminal one is staminate.

COMMENTS: *Carex serratodens* is fairly rare in OR, though more common on serpentine substrates in northern CA. The roots form a large, dense, soil-holding mass, suggesting this species is valuable for preventing soil erosion. It is available commercially for use in gardens. The dense clumps can be impressive, but the culms tend to flop over and look somewhat messy. The main advantage of *C. serratodens* as a garden plant is its ability to grow in partial shade and in difficult substrate including serpentine and clay.

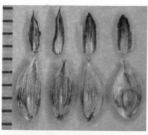

Carex serratodens

top left: pistillate scales, perigynia
top right and center left : inflorescences
bottom left: inflorescences, habit
bottom right: gynecandrous terminal
    spike

Carex sheldonii Mack.
Common name: Sheldon's Sedge
Section: *Carex*   Key: C

KEY FEATURES:
- Perigynia with dense short hairs and short, spreading teeth
- Leaves and summit of leaf sheath front hairy
- Riparian wetlands E of Cascades

DESCRIPTION: Habit: Rhizomatous, forming large populations. **Culms**: Fertile culms 40-80 cm tall. **Leaves**: Usually spreading hairy (occasionally glabrous), 2.5-6 mm wide. Basal leaf sheaths reddish purple or drab. Leaf sheath front pubescent at least near the top, becoming ladder-fibrillose with age. **Inflorescences**: Lateral (1-) 2-3 spikes ♀, 2-6 cm long, erect or ascending. Terminal 1-4 spikes ♂, 1.5-4.5 cm long. ♂ **scales**: Awnless, glabrous or sparsely appressed-pubescent near the tip. ♀ **scales**: Narrow, scabrous along midvein but not hairy, with scabrous awn. **Perigynia**: Sparsely to densely pubescent, lanceolate or lance-ovate, with 12-18 veins, 4.8-6.5 mm long, 1.4-2.4 mm wide. Beak 1.4-2.6 mm long, with spreading teeth (0.4-) 0.6-1.4 mm long. Stigmas 3, style persistent. **Achenes**: Trigonous.

HABITAT AND DISTRIBUTION: Wet meadows, ditch banks, riparian meadows, openings in riparian forest, on stable fluvial surfaces above the annual flood line, often in somewhat alkaline sites; E OR. OR, ID, NV, and rare in NE CA.

IDENTIFICATION TIPS: *Carex sheldonii* is rhizomatous, sometimes forming dense stands. Its more or less pubescent leaf blades and leaf sheath fronts and its pubescent, strongly beaked perigynia make this species easy to identify. It is similar to the coarser *C. atherodes,* which grows in wetter conditions, sometimes adjacent to *C. sheldonii. Carex atherodes* has wider leaves, longer perigynia, longer beak teeth, and sparsely hairy perigynia. *Carex pellita* may grow with *C. sheldonii,* but its foliage is glabrous and its perigynia are shorter (2.4 – 5.2 mm long). *Carex hirta,* introduced at Portland, has ♂ scales with dense, spreading, white hairs.

COMMENTS: *Carex sheldonii* is much appreciated by the botanist because it is easily identified. Like most riparian sedges, it can be harmed by heavy grazing along riparian zones. This species produces a remarkably dense mass of rhizomes and roots. It is an effective soil binder that would be useful for habitat restoration along somewhat alkaline stream terraces. Unlike most *Carex,* this species and *C. atherodes* have true vegetative culms, with nodes above ground level.

Carex sheldonii

top left: pistillate scales, perigynia, achenes
top middle, left: hairy leaf sheath
top middle, right: ladder-fibrillose leaf sheath
top right: spike
bottom: inflorescence

355

Carex siccata Dewey
Common name: Dry-spike Sedge
Section: *Ammoglochin*   Key: main, H
Synonyms: *C. foenea* Willd., misapplied. *C. aenea*
   Fernald, misapplied, *C. foenea* var. *tuberculata*
   F. J. Hermann

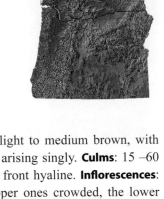

KEY FEATURES:
• Rhizomes with pithy cortex
• Long perigynium beak with flat margins
• Dry upland grassland or open forest

DESCRIPTION: Habit: Rhizomatous, the rhizomes light to medium brown, with a pithy cortex that can be peeled off, most shoots arising singly. **Culms**: 15 –60 (-90) cm tall. **Leaves**: 1-3.2 mm wide; leaf sheath front hyaline. **Inflorescences**: 1-5 cm long, with 4-12 ascending spikes, the upper ones crowded, the lower sometimes more separated. Lower spikes ♀, middle ones often ♂. Terminal spike ♀ or androgynous (sometimes apparently gynecandrous because it is ♀ and right below it are 2 or more small ♂ spikes). **♀ scales**: Reddish brown with pale or green center and hyaline margins, shorter than or nearly as long as the perigynia, with apex acute to acuminate. **Perigynia**: Lanceolate to elliptic, with several dorsal veins, sometimes veinless on the ventral surface, sometimes with many small bumps on the ventral surface, (3.9-) 4.5-6.5 mm long, 1.4-2.4 mm wide, with narrow wings on upper body and/or lower part of beak, with relatively long beak 1.2-2.7 mm long, more than half as long as or sometimes equal to the perigynium body, serrulate margined. Stigmas 2. **Achenes**: Lenticular.

HABITAT AND DISTRIBUTION: Open conifer forest, upland grasslands and savanna; Chelan and Kittitas cos., WA. YT to ME, S to WA, AZ, NM, IL, and NJ.

IDENTIFICATION TIPS: Grass-like *Carex siccata* often mixes inconspicuously with other graminoids but may dominate mesic swales. *Carex praegracilis* is superficially similar but has blackish rhizomes and shorter perigynia. *Carex sartwellii*, found in BC but not OR or WA, has green leaf sheath fronts and smaller perigynia.

COMMENTS: *Carex siccata* can be a community dominant in the herbaceous layer of open conifer forest. Where common, it can furnish good forage for cattle and horses, but it is rare in the PNW. It reduces erosion, especially in sandy soils, and has been used in habitat restoration projects outside the PNW, sometimes on old mine sites. This species has been involved in a three-way confusion of names involving *C. foenea* and *C. aenea*. *Carex siccata* is the only one of the three actually known to grow in the PNW.

top left: perigynium
top middle and right: inflorescences
center left and bottom: habit

Carex simulata Mack.
Common name: Short-beak Sedge
Section: *Divisae*   Key: B, H

KEY FEATURES:
- Perigynia short-beaked, dark brown, shiny
- Separate ♂ and ♀ plants
- Dense, monospecific stands in wet meadows E of Cascades

DESCRIPTION: Habit: Rhizomatous, the rhizomes pale to brown, 1.5-2.8 mm in diameter; forming large dense stands. **Culms**: (10) 20-50 (-100) cm. Shoot bases brownish. **Leaves**: 1-3.7 mm wide. **Inflorescences**: 1-3 cm long, of several small spikes. Plants apparently dioecious; flowers in one plant seemingly all ♀ or all ♂ but usually with a few, nearly undetectable flowers of the opposite sex mixed in the spikes. ♀ **scales**: Reddish brown with hyaline margins. **Perigynia**: Very dark brown, glossy, stout, 1.8-2.8 mm long, 1.1-1.7 mm wide. Beak 0.25-0.5 mm long. Stigmas 2. **Achenes**: Lenticular. **Anthers:** With a tiny smooth to warty awn.

HABITAT AND DISTRIBUTION: Wet meadows and springs, on stable stream terraces that are not scoured, where the water table is at or above the soil surface until late summer and even in August the soil is saturated just below the surface, generally on calcareous or slightly alkaline substrates; Cascades and widespread E of Cascades. S BC to SK, south to CA and NM.

IDENTIFICATION TIPS: In a dense sward of narrow-leaved sedges in a moist habitat E of the Cascade crest, knock off a few perigynia into your hand. If perigynia are short, plump, and dark brown, it's *C. simulata*. The males can be very difficult to understand the first time one encounters them, and are best identified by the presence of the female clones nearby. Sometimes a male clone becomes established alone; the narrow leaves, rhizomatous habit, dull plant bases, and habitat make identification possible. *Carex praegracilis* and other related dioecious species grow in drier habitats, are shorter, and have longer perigynium beaks. *Carex praegracilis* has blackish rhizomes and shoot bases.

COMMENTS: *Carex simulata* can form extensive, fine-textured, single-species stands that stabilize soil in wet meadows, headwaters, and swales. Male and female clones may occupy different parts of the stand. Ungulates eat the foliage readily, but it is usually too sporadic in occurrence to contribute a large proportion of their diet.

Carex simulata

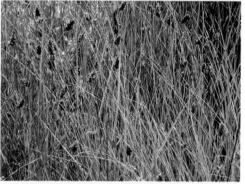

top left: pistillate scales, perigynia
top right: staminate and pistillate
   inflorescences
center: habit
bottom: habitat

Carex spectabilis Dewey
Common name: Showy Sedge
Section: *Scitae*   Key: F
Synonyms: *C. tolmiei* Boott

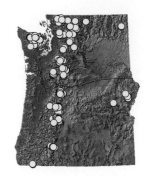

KEY FEATURES:
• Forming clumps from short rhizomes
• Upper spikes erect, lower ones drooping
• Sunny, subalpine to alpine habitat

DESCRIPTION: Habit: Loosely cespitose to short rhizomatous. **Culms**: (10-) 25-50 cm tall, aphyllopodic. **Leaves**: 2-5 mm wide, slightly glaucous. **Inflorescences**: (2-) 3-5 (-9) spikes, each 0.8-2 cm long. Lateral spikes ♀, the upper ones erect and the lower ones drooping, often on long peduncles. Terminal spike ♂. ♀ **scales**: Brown or black with dark margins, midvein light and conspicuous (rarely black), narrower and shorter than (sometimes as long as) the perigynia, with midrib sometimes extended as a short awn to 1 mm long. **Perigynia**: Greenish or purplish black, ovate, flattened, veined, with the marginal ribs at the edges (or sometimes displaced away from the margin), 3.5-5 mm long, 1.75-2 mm wide, with beak 0.4-0.5 mm long. Stigmas 3. **Achenes**: Trigonous, filling half or less of the perigynium bodies.

HABITAT AND DISTRIBUTION: Subalpine and alpine meadows and roadsides, generally in well-drained soils but on sites that are moist due to high precipitation, snowmelt, or location in a depression; on slopes where snow melts relatively early (in June or July); widespread in mts of the PNW, but apparently not on Steens Mt. AK to SK, S to CA and UT.

IDENTIFICATION TIPS: This frustratingly variable species can form lush, crowded mounds of slightly glaucous leaves, but on drier slopes it may produce scattered, dwarfed shoots with 1 or 2 spikes. Plants are often collected immature (even in late summer), when the halo of yellow anthers and the white stigmas contrast with the dark scales. Very similar *C. paysonis* has blades on the lower leaves of its flowering culms, and broader, rounder perigynia with the marginal veins consistently displaced away from the margins. Other similar species differ because they have gynecandrous terminal spikes. See *C. heteroneura* and *C. raynoldsii*.

COMMENTS: This species is often a community dominant in subalpine or alpine meadows. It is an important food for wildlife and it is used for habitat restoration projects. Recent molecular phylogenetic studies suggest that, although they have trigonous achenes, *C. spectabilis* and the other sedges in section *Scitae* belong in section *Phacocystis* with *C. scopulorum* and relatives. Trying to distinguish *C. spectabilis* from *C. paysonis* may drive you nuts.

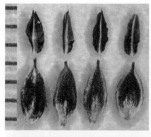

top left: pistillate scales, perignia
top right and center left:
    inflorescences
bottom: habit, habitat

Carex stipata Muhl. ex Willd. var. *stipata*
Common name: Awl-fruit Sedge
Section: *Vulpinae*   Key: H

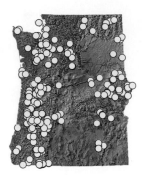

KEY FEATURES:
- Leaf sheath fronts cross-corrugated
- Large, lance-triangular perigynia with swollen, pithy bases
- Culms winged, spongy, sharply triangular

DESCRIPTION: Habit: Densely cespitose. **Culms**: 30-120 cm tall, thick, winged, and spongy. **Leaves**: 5-11 mm wide. Leaf sheaths with blades, the fronts cross-corrugated, not veined, fragile, hyaline, with concave mouth. **Inflorescences**: Dense heads (2-) 3-10 cm long, 1-3 cm wide, with androgynous spikes (or some all ♀). Beaks of the spreading perigynia give the inflorescence a prickly look. ♀ **scales**: Hyaline to brownish, shorter than the perigynia. **Perigynia**: Lance-triangular, widest near the base, green, maturing yellowish to pale brown, (3.6-) 4-5.2 mm long, about 2 mm wide, strongly veined, with 15 reddish brown veins on the dorsal surface, 7 on the ventral surface, the base pithy, swollen, cordate, tapering gradually to a poorly defined beak that is about 2.5 mm long. Stigmas 2. **Achenes**: Lenticular.

HABITAT AND DISTRIBUTION: Marshes, edges of thickets, and other wet spots with still water, usually in full sun, sometimes in shade, at low to moderate elevations; throughout WA and OR. AK to NF, S to CA, TX, and SC, also East Asia.

IDENTIFICATION TIPS: *Carex stipata* is densely cespitose with a large, spiky inflorescence, large, green to straw-colored perigynia, and cross-corrugated leaf sheath fronts. *Carex densa* and *C. vulpinoidea* grow in the same habitat, often adjacent to *C. stipata*. They have smaller perigynia in narrower inflorescences, and they lack winged culms. The large perigynia of *C. stipata* might suggest *C. kobomugi* or *C. macrocephala,* but those are tough, strongly rhizomatous, dioecious species of sandy soils, and they lack cross-corrugated leaf sheath fronts. Runty *C. stipata* can be confused with *C. neurophora,* which has smaller perigynia that are 2.9-3.8 mm long.

COMMENTS: *Carex stipata* grows in wetlands, preferring wet marsh edges that are seasonally flooded. It produces relatively large amounts of seed. The pithy tissue at the base helps the perigynia float well, facilitating water-borne seed dispersal and making them attractive to waterfowl as food. It is an early-successional plant used for habitat restoration, wetland mitigation projects, and stormwater control basins at low to moderate elevations throughout our region, where water is still or slow moving.

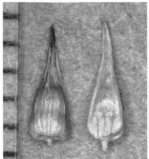

top left: perigynia
top middle and top right:
   inflorescences
bottom: habit

Carex straminiformis Mack.
Common name: Shasta Sedge
Section: *Ovales*   Key: J

KEY FEATURES:
- Cespitose, with gynecandrous spikes and winged perigynia
- Wide, flat perigynia with very broad, often crinkled margins
- Plants short, with leaves usually folded
- Dry alpine ridges, pumice plains, lake terraces

DESCRIPTION: Habit: Cespitose. **Culms**: 20-50 cm. **Leaves**: 8-25 cm long, 3-4 mm wide, flat or folded. **Inflorescences**: 1.3-3 cm long, 12-20 mm wide, the 3-8 spikes crowded but distinguishable. Spikes gynecandrous. ♀ **scales**: White to reddish brown with pale midstripe, lanceolate to ovate, 3.2-4.8 mm long, shorter and narrower than the perigynia, the tip acute to acuminate. **Perigynia**: Straw-colored, green, or whitish with brown tip, broadly ovate, flat, 4-5.8 mm long, 1.8-3.4 mm wide, widest below middle of total perigynium length, with 10-20 dorsal veins and 1-12 ventral veins, with wings 0.4-1 mm wide, the edges often crinkled. Beak winged and ciliate-serrulate to near the tip (or sometimes unwinged, brown, parallel-sided, and entire for the distal 0.5-0.7 mm); 2.2-3 mm from achene top to beak tip. Stigmas 2. **Achenes**: Lenticular, located below the center of the perigynium, (1.4-) 1.7-2.4 mm long, 1-1.6 mm wide, 0.3-0.5 mm thick.

HABITAT AND DISTRIBUTION: In rocky or gravelly soils, dry open slopes, sometimes in open forests, at or near timberline, often near persistent snowbanks, mainly in the high Cascades of WA and OR, but also at Mt. Ashland, Newberry Crater, and the Warner Mts. in OR and to be looked for elsewhere. WA to MT, S to CA and UT.

IDENTIFICATION TIPS: *Carex straminiformis* is a coarse, tough sedge with wide, flat, perigynia. The wings are often crinkled distally. The other high-elevation sedge with broad, flat, "potato chip" perigynia is *C. proposita*, known in WA but not OR. Its perigynia are widest at about the middle (of total perigynium length) and the achene is in the center of the perigynium. Its perigynium abruptly narrows to a short beak. *Carex proposita* is shorter on average, 10-35 cm tall, with narrower (1-2 mm wide) leaves that are usually folded. Although *C. brevior* lives in mesic sites at low to moderate elevations, it has been confused with montane *C. straminiformis*. Its perigynia are very broad but planoconvex, and its beak is shorter and broader.

COMMENTS: *Carex straminiformis* can be a community dominant in its rocky, high-elevation habitat.

top left: perigynia
top right: inflorescence
bottom right: habit

Carex stylosa C. A. Mey.
Common name: Long-style Sedge
Section: *Racemosae*   Key: F

KEY FEATURES:
• Ascending spikes with blackish scales
• Cespitose
• Forming clumps in bogs in NW WA

DESCRIPTION: Habit: Loosely cespitose. **Culms**: 15-50 cm tall. **Leaves**: 2-4 mm wide. **Inflorescences**: (2-) 3-4 (-5) spikes, each 0.7-2 cm long, the lateral spikes ♀ and erect on short peduncles, the terminal spike ♂ or occasionally with a mix of ♂ and ♀ flowers. ♀ **scales**: Purplish black or dark brown with hyaline margins, equaling or a little shorter than the perigynia, with a lighter midvein. **Perigynia**: Spreading, green or yellow brown, becoming brown, or sometimes dark like the scales, elliptic, 2.5-3.5 mm long, 1.5-1.75 mm wide, veinless except for the two marginal ribs, beakless or with a short (0.2-0.3 mm long) abrupt beak. Style persistent and sticking out of the perigynium for a while after the 3 stigmas have fallen, but eventually deciduous. **Achene**: Trigonous, nearly filling the perigynia.

HABITAT AND DISTRIBUTION: High-elevation bogs in mountains of NW WA; elsewhere, marshes, wet meadows, fens, bogs, streambanks, sometimes on gravelly soils or among rocks, from sea level to the alpine zone. AK to WA, disjunct QC to NF, Greenland; East Asia including Japan.

IDENTIFICATION TIPS: *Carex stylosa* forms large spreading clumps in acid bogs, and has dark pistillate scales. It can be confused with the much more common *C. scopulorum,* which is more strongly rhizomatous, occurs in less soggy-wet habitats, lacks the long, persistent style, and normally has 2 stigmas and lenticular achenes. Very rarely, confusing individuals of *C. scopulorum* have 3 stigmas and trigonous achenes.

COMMENTS: Like *C. anthoxanthea, C. buxbaumii, C. livida, C. macrochaeta, C. media,* and several others, *C. stylosa* has been retreating N with the glaciers, leaving small, isolated populations behind. These isolated populations are interesting evolutionary experiments. Their range of genetic variation differs from that of the main populations. This is obvious in the case of *C. macrochaeta,* which has much shorter awns in the PNW than do northern plants. Isolated populations may continue changing and produce descendents so different from their widespread relatives that we will call them distinct species. Preserving these ice age relics will become increasingly difficult as the climate continues to warm.

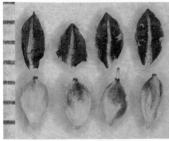

top left: pistillate scales, perigynia
top right: inflorescences
bottom: habit, habitat
  (*Sphagnum* bog)

Carex subbracteata Mackenzie
Common name: Smallbract Sedge
Section: Ovales　　　　　Key: J

KEY FEATURES:
- Cespitose, with gynecandrous spikes & winged perigynia
- Thick, coppery, ascending perigynia
- Near the coast

DESCRIPTION: Habit: Cespitose. **Culms**: 25–105 cm tall, sometimes lax and leaning. **Leaves**: 1.3-4.6 mm wide, sheath sometimes extending above collar. **Inflorescences**: 1.3-3.5 cm long, 7-23 mm wide, dense and head-like, (often elongate on late season shoots), lowest internode 2-3(-4.5) mm long, generally shorter than lowest spike, inflorescence bract usually short and bristle-like, sometimes leaf-like and longer than the inflorescence. **Spikes:** Gynecandrous. ♀ **scales**: Brown, red-brown, or coppery, with whitish, green, or pale brown midstripe, 3.5-4.5 mm long, usually shorter and narrower than the perigynia, sometimes longer and covering them, margin white, 0-0.4 mm wide. **Perigynia**: Usually coppery with brown margin and a metallic sheen, ascending, ovate to broadly ovate, planoconvex, (2.9-)3.5-4.7 (-5.7) mm long, usually 1.3-1.7(-2.2) mm wide, 0.5-0.7 mm thick, leathery; with (0-)5-9 conspicuous veins dorsally, 0-7 conspicuous veins ventrally, the ventral veins shorter than or as long as the achenes; wing 0.2-0.3(-0.4) mm wide; margin ciliate-serrulate above. Base rounded to bluntly angled, without a stipe. Beak +/- 1.5-2.5 mm from achene top to beak tip, tip brown or white, cylindric, unwinged, and +/- entire for 0.4-0.8 mm. Stigmas 2. **Achenes**: Lenticular, (1.3)1.5-2.1 mm long, 0.9-1.7 mm wide, 0.5-0.6 mm thick.

HABITAT AND DISTRIBUTION: Seasonally moist soil, grasslands to open forest, coastal SW OR to CA, introduced along the coast to central OR, WA, and BC.

IDENTIFICATION TIPS: The much more common *C. pachystachya* has more spreading perigynia that make the mature spikes star-shaped in outline, and pistillate scales that do not cover the perigynia and are dark throughout or have white margins less than 0.1 mm wide. Very similar *C. harfordii* has ventral perigynium veins usually reaching the base of the beak. Late-season *C. subbracteata* shoots with leaf-like inflorescence bracts might be confused with paler *C. athrostachya*.

COMMENTS: *Carex subbracteata* is similar to and perhaps conspecific with *C. gracilior*, a CA endemic with more open inflorescences. Sterile putative hybrids with similar C. harfordii have been reported from coastal S CA.

Carex subbracteata

top left: perigynia
top right, center right, below left:
 inflorescences
bottom right: habitat (coastal
 hills)

Carex subfusca W. Boott

Common name: Pale Broom Sedge
Section: *Ovales*   Key: J
Synonyms: *C. teneriformis* Mackenzie

KEY FEATURES:
- Cespitose, with gynecandrous spikes and winged perigynia
- Inflorescence fine textured, pale green to straw-colored, with overlapping spikes
- Perigynia small, with narrow wings

DESCRIPTION: Habit: Cespitose. **Culms**: 20-105 cm tall. **Leaves**: 6-45 cm long, 1.2-3 (-3.7) mm wide. Leaf sheath front white-hyaline. **Inflorescences**: 1.1-3 cm long, 5.4-16.3 mm wide, usually green to straw-colored (sometimes brownish in northern plants), spikes overlapping but more or less distinguishable, often crowded, fine-textured because the perigynia are small. Spikes gynecandrous. ♀ **scales**: White-hyaline to gold, occasionally brown, 2.1-3.5 (-3.9) mm long, shorter than or more or less covering the perigynia. **Perigynia**: Green, whitish, straw-colored, or light brown, lance-ovate to ovate, 2.4-3.5 mm long, 0.9-1.9 mm wide, with (0-) 2-11 dorsal veins and 0-6 ventral veins. Beak tip winged and ciliate-serrulate nearly to the tip, or unwinged, parallel-sided, and more or less entire for the distal 0.4-0.7 mm; 1.2-2 mm from top of achene to tip of beak. Stigmas 2. **Achenes**: Lenticular, 1-1.6 mm long, 0.7-1.25 mm wide.

HABITAT AND DISTRIBUTION: Scattered small populations in mesic to moist meadows, seeps, roadsides, and lakeshores, sometimes in moderately disturbed sites, usually in the sun but sometimes in partial shade, mainly in mts; widespread in OR, also S WA. WA S to CA, NM, and Mexico, reported N to Yukon. Reported in Hawaii but plants there are atypical.

IDENTIFICATION TIPS: *Carex subfusca* is identified by its pale to greenish, fine-textured, erect, somewhat elongated inflorescences and its small achenes. It is confused with most other *Carex* section *Ovales*, particularly with individuals that have small perigynia. *Carex feta* and *C. fracta* have longer inflorescences. In addition, *C. feta* has green leaf sheath fronts. Like *C. subfusca*, *C. bebbii* has fine-textured inflorescences, but it is distinguished by its brown ♀ scales and perigynia, and its rounder spikes.

COMMENTS: Diverse genetic markers and relatively great morphological variation suggest that two or more species hide under the name *C. subfusca*, and the taxonomy of this complex needs more study. Because of this uncertainty, *C. subfusca* seed used for habitat restoration projects should only be collected from areas near the planting sites.

top left: pistillate scales, perigynia
top right and bottom right: inflorescences
bottom left: habit

Carex subnigricans Stacey
Common name: Nearlyblack Sedge
Section: *Inflatae*   Key: A

KEY FEATURES:
- Single, thick spike
- Oval, flat perigynia
- High elevation, moist, rocky slopes

DESCRIPTION: Habit: Rhizomatous, the rhizomes 1 mm thick. **Culms**: 5-20 cm tall. **Leaves**: Round to semicircular in cross section, 0.4-1.0 mm wide, the leaf sheath fronts uniformly pale brown or colorless. **Inflorescences**: One androgynous spike per culm, 0.7-2 cm long, 4-6 mm wide, lacking inflorescence bract. ♀ **scales**: About as wide as the perigynia but shorter, 1-veined, yellowish brown, obtuse to acute and flat at tip, not awned. **Perigynia**: Ascending, brown above, darker brown through the middle, and pale below, elliptic, veinless, 2.5-3.5 mm long, 1.5 mm wide, with rachilla shorter than the achene and the achene much smaller than the perigynium. Stigmas 3. **Achenes**: Trigonous.

HABITAT AND DISTRIBUTION: Moist rocky slopes and meadows, alpine to subalpine zones; rare in the Wallowas and Steens Mt., OR; mts of the intermountain region; OR to WY, S to central CA, NV, and UT.

IDENTIFICATION TIPS: *Carex subnigricans* is a short, turf-forming, high elevation species. Each flowering culm produces a single, oval spike. Superficially it is very similar to young *C. nigricans* except that *C. nigricans* has flat leaves 1.5-4 mm wide. At maturity, the broadly lanceolate perigynia of *C. nigricans* reflex and the ♀ scales fall off. *Carex engelmannii* and *C. breweri* have broader heads and live in drier habitats. In addition, *C. emgelmannii* has rounded, sessile perigynium bases, and ♀ scales that appear acuminate. *Carex breweri* has 3-5 veins on the pistillate scales and grows in the Cascade/Sierra Nevada axis.

COMMENTS: In *C. subnigricans* and its relatives, the leaves are narrow and somewhat rolled up, reducing the surface area exposed to the air and therefore conserving water. The perigynium encloses a rachilla, a vestigial branch. A microscopic version of this branch is probably present in all *Carex* and it may become elongated. For example, rare plants of *C. aquatilis* var. *dives, C. obnupta,* and *C. hassei* have a rachilla with bracts sticking out of the tip of the perigynium. In some cases, a whole spike may develop from an elaborated rachilla; a lateral spike in *C. aurea* may have a perigynium-like bract at its base. The rachilla in *Carex* and the short branch that supports the staminate flower within the *Kobresia* perigynium evolved from the same structure in their common ancestor.

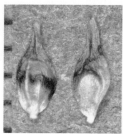

top left: perigynia
top right: inflorescences
bottom: habit

Carex sychnocephala J. Carey
Common name: Many-headed Sedge
Section: *Ovales*   Key: I, J

KEY FEATURES:
- Cespitose, with gynecandrous spikes and narrowly winged perigynia
- Lower inflorescence bracts leaf-like, 5+ times as long as inflorescence
- Perigynium beak much longer than body
- Disturbed and seasonally wet areas

DESCRIPTION: Habit: Cespitose. **Culms**: 8-40 cm tall. **Leaves**: 12 cm long, 1.2-3 mm wide. Leaf sheath fronts white-hyaline. **Inflorescences**: 1.6-3 cm long, 7-15 mm wide, the spikes crowded in a dense greenish to golden head, the lower inflorescence bracts leaf-like and usually at least 5 times as long as the inflorescence. Spikes gynecandrous. ♀ **scales**: White- or gold-hyaline, 3.5-4.5 mm long, shorter or longer than the perigynia. **Perigynia**: Green, straw-colored, or light brown, narrowly lanceolate, (4.6-) 5.5-7.3 mm long, 0.7-1.2 mm wide, with wing 0.1 mm wide on the perigynium body, with 3-12 dorsal veins, ventrally veinless or with up to 9 veins. Beak winged and ciliate-serrulate to near the tip; 3-5 mm from achene top to beak tip. Stigmas 2. **Achenes**: Lenticular, 1-1.8 mm long, 0.6-0.8 mm wide, 0.3-0.4 mm thick.

HABITAT AND DISTRIBUTION: Open, seasonally wet shores of ponds and rivers; rare in NE WA and Harney Co., OR. AK to QC, S to OR, MO, and NY, also CO.

IDENTIFICATION TIPS: *Carex sychnocephala* is unmistakable with its extremely long inflorescence bracts, which resemble the leaves, and its long, narrow perigynia that are more beak than body. *Carex athrostachya* has elongated inflorescence bracts and may occur in similar habitats, but its bracts and lanceolate perigynia are much shorter.

COMMENTS: Like *C. athrostachya* and *C. bebbii*, *C. sychnocephala* is a short-lived perennial that can bloom and set seed in its first year. It may die the first year, and thus may function as an annual. This is a good adaptation for a species of shifting river channels and pond margins. *Carex sychnocephala* is rare in the PNW. The main threats here are hydrological alteration of its habitat and recreational development. It can withstand some grazing and trampling, but heavy grazing can be a problem at some sites. It has such odd perigynia that it has been placed in its own section, *Cyperoideae*, together with its Eurasian counterpart, *C. bohemia*. However, recent DNA-based studies indicate that these species are aberrant members of section *Ovales*.

top left: pistillate scales, perigynia
top right: inflorescence
center left: habit
botttom: inflorescences, habitat

Carex sylvatica Huds.
Common name: European Woodland Sedge
Section: *Hymenochlaenae* (or *Porocystis*?)   Key: F

KEY FEATURES:
- Long, dangly ♀ spikes
- Stigmas 3, perigynia trigonous
- Moist forests with clay soils

DESCRIPTION: Habit: Cespitose. **Culms**: 25-110 (-200) cm tall, erect, much longer than the leaves, phyllopodic. **Leaves**: Arching, (3-) 5.5-8.5 (-15) mm wide, glabrous, with finely scabrous margins. Lowest leaf sheaths off-white or brown. Leaf sheath fronts whitish hyaline. **Inflorescences**: Inflorescence bracts with well developed sheaths. Lateral spikes 3-5, ♀ (or the upper ones androgynous, sometimes even ♂), cylindrical, 1.5-6 cm long, 3-5 mm wide; the lower ones dangling, the upper ones sometimes erect. Terminal spike ♂ (or sometimes with a few perigynia near the base), 1.5-4 cm long, on peduncle less than 2 cm long. ♀ **scales**: White-hyaline with green midstripe, shorter than the perigynia. **Perigynia**: Glabrous, green (maturing brown), with 2 marginal ribs but otherwise veinless (or with short veins at base), 4.5-6 mm long, 1.4-1.8 mm wide, narrowed to a stalk-like base, the body obovoid, narrowed to the prominent, 2-3 mm long beak, which is nearly half of the perigynium length. Stigmas 3. **Achenes**: Trigonous.

HABITAT AND DISTRIBUTION: Disturbed areas in moist forest; introduced to Seattle, WA, and islands in the Gulf of Georgia, SW BC. Native to Europe, introduced to eastern N America (ON, NY, NC) and New Zealand.

IDENTIFICATION TIPS: The green to brown, dangly spikes and long-beaked perigynia make fertile *C. sylvatica* different from any other PNW sedge. *Carex leptopoda* and its relatives occupy similar habitat but have softer leaves, shorter spikes, and perigynia tapering to shorter beaks. The disturbed forest habitat and tough, arching, green leaves of sterile plants might suggest *C. rossii*, but that species has reddish shoot bases and short inflorescences with pubescent perigynia. *Carex sprengelii* (not in WA or OR but in BC and MT) has long tubular beaks somewhat like those of *C. sylvatica,* but it is short-rhizomatous, its shoots have dense, brown fibrous remains of previous year's leaves at the base, and it has shorter (10-35 mm), broader (8-10 mm) ♀ spikes.

COMMENTS: In Europe, *C. sylvatica* is often a community dominant in forests on alluvial clay soils. It is planted in gardens and has escaped in several places in N America. It creates a persistent seed bank in the soil. Because it is mostly self-pollinating, one established plant can produce a whole population.

top left: pistillate scales, perigynia
top right: inflorescence
bottom: habit, habitat

Carex tahoensis Smiley
Common name: Tahoe Hare Sedge
Section: *Ovales*   Key: J
Synonyms: *C. phaeocephala* Piper, misapplied

KEY FEATURES:
- Cespitose, with gynecandrous spikes and winged perigynia
- Spikes pointing stiffly upward, overlapping, in a narrow inflorescence
- Perigynia opaque, brown, and strongly veined
- Dry, subalpine slopes and grasslands

DESCRIPTION: Habit: Cespitose. **Culms**: 15-45 cm tall. **Leaves**: 5-10 cm long, 1.5-2.5 (-3) mm wide, folded, somewhat tough, the edges sometimes revolute. **Inflorescences**: (1.5-) 2-3 (-3.7) cm long, 6-12 mm wide, somewhat elongated, erect or somewhat angled to one side. Spikes gynecandrous, narrow, ascending, overlapping. ♀ **scales**: Gold to brown with paler midrib, 4-5 mm long, more or less covering the perigynia. **Perigynia**: Dark reddish brown with green or pale edges, opaque, and somewhat tough when mature (when very immature, pale, translucent, and with inconspicuous veins), ovate, (3.7-) 4.5-6 mm long, 1.5-2.6 mm wide, 0.5-0.7 (-0.9) mm thick, strongly 3-8 veined on the ventral surface, with 7-15 strong veins on the dorsal surface. Beak tip unwinged, brown, parallel-sided, and entire for the distal 0.2-0.7 mm, with the perigynium wings extending up to the base of the cylindric tip, giving the perigynium a somewhat boat-like shape; (0.6-) 1-2 mm from beak tip to achene top. Stigmas 2. **Achenes**: Lenticular, 1.9-2.4 mm long, 1.2-1.6 mm wide, 0.5-0.7 mm thick.

HABITAT AND DISTRIBUTION: Dry subalpine and alpine ridges and grasslands; Olympics and N Cascades, WA, mountains of E OR. AK to AB, S to CA and CO.

IDENTIFICATION TIPS: *Carex tahoensis* is a tough, cespitose, subalpine sedge with perigynia ascending to appressed, opaque, and mostly hidden by the ♀ scales. *Carex phaeocephala* is superficially very similar but its perigynia are ± translucent, usually paler, veinless (or with few faint veins), and shorter, narrower, and particularly less thick. Plants with immature perigynia cannot be distinguished. *Carex leporinella* grows in wet sites and has shorter perigynia (3.5-4.2 mm long). *Carex petasata* is typically taller with longer perigynia, and grows at lower elevations.

COMMENTS: *Carex tahoensis* is well suited to revegetation projects on dry, rocky, subalpine roadsides and recreation sites. Because of taxonomic confusion, reports on ecology or use of *C. phaeocephala* may include information on *C. tahoensis*. The full range of *C. tahoensis* is not known.

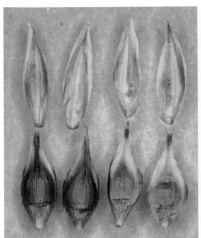

top left: pistillate scales, perigynia   *top right*: inflorescences (with some
   smutty perigynia)
center: habit   *bottom*: habitat

Carex tenera Dewey var. *tenera*
Common name: Tender Sedge
Section: *Ovales*   Key: J

KEY FEATURES:
- Cespitose, with gynecandrous spikes and winged perigynia
- Inflorescence typically nodding, with at least the lower spikes separated
- Perigynia ovate and flat

DESCRIPTION: Habit: Cespitose. **Culms**: 20-90 cm tall. **Leaves**: 15-35 cm long, 1.3-2.5 (-3) mm wide. Leaf sheaths smooth or minutely papillose near the top, the fronts white-hyaline, the backs mottled with white. **Inflorescences**: (2-) 2.5-5 cm long, 7-10 mm wide, usually nodding with a delicate, hair-like axis, the spikes distinct. Spikes gynecandrous. ♀ **scales**: White hyaline or pale brown with green or brown midstripe, 2.3-3.3 mm long, narrower than and as long as or shorter than the perigynia; tip obtuse on lower scales, acute on upper scales. **Perigynia**: Brown, ovate, 2.8-4 (-4.5) mm long, 1.4-1.9 (-2) mm wide, wings 0.1-0.5 mm wide, with 5-7 dorsal veins and 0-7 ventral veins. Beak winged and ciliate-serrulate almost to the tip; 1.3-2.7 mm from achene top to beak tip. Stigmas 2. **Achenes**: Lenticular, 1.3-1.7 mm long, 0.8-1.1 mm wide, 0.5 mm thick.

HABITAT AND DISTRIBUTION: Meadows, thickets, and riparian forest openings, in moist to mesic areas in sun or partial shade; Okanogan and Pend Oreille cos., WA. BC to ME, S to WA, WY, and VA.

IDENTIFICATION TIPS: *Carex tenera* is usually characterized by its elongated, nodding inflorescence with a very narrow, almost hair-like axis. A more subtle clue is provided by the vegetative culms, which have a few leaves clustered near the tips; most PNW species have leaves crowded near the base or more evenly spaced. *Carex scoparia* plants with elongated inflorescences can resemble *C. tenera,* but spikes are green overall and very fine textured due in part to the acuminate ♀ scales. *Carex tenera* with condensed inflorescences are most likely to be confused with *C. brevior,* but the latter's perigynia lack veins or have only 1-5 faint or irregular dorsal veins.

COMMENTS: In eastern N America, where it is common, *C. tenera* has been used in wetland restoration projects and rain gardens, particularly in partial shade. Eastern seed should not be planted in the PNW. Here, the few populations of *C. tenera* should probably be left undisturbed, or PNW seed could be used in carefully designed restoration projects within its very limited native range.

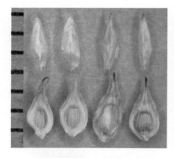 

top left: pistillate scales, perigynia
top right: inflorescence
botttom: habit

Carex tenuiflora Wahlenb.
Common name: Sparse-flower Sedge
Section: *Glareosae*   Key: I

KEY FEATURES:
- Few small spikes crowded at tip of culm
- Beakless, green perigynia
- Delicate plant of moss mats in bogs

DESCRIPTION: Habit: Loosely cespitose or rhizomatous, with slender rhizomes. **Culms**: 10-50 cm long, more or less erect but sometimes weak. **Leaves**: Green or yellowish green, flat or channeled, 0.5-2 mm wide. **Inflorescences**: 0.6-1.2 cm long, with 2-4 (usually 3) gynecandrous, subglobose spikes, each with 3-15 perigynia, crowded at the tip of the culm. ♀ **scales**: White-hyaline marked with green, about as long as the perigynia. **Perigynia**: Appressed-ascending, green or gray-green, obovate to elliptic, with a few obscure veins, 3-3.5 mm long and 1.5-1.8 mm wide, beakless or nearly so. Stigmas 2. **Achenes**: Lenticular.

HABITAT AND DISTRIBUTION:*Sphagnum* bogs, fens, cedar swamps, often in partial shade; 1000-5000 feet; Okanogan Co., WA. AK to NF, S to WA and ME, northern Eurasia.

IDENTIFICATION TIPS: This slender, green bog plant superficially resembles *C. leptalea,* which has a single, androgynous spike and trigonous achenes. *Carex canescens* can grow with *C. tenuiflora*. It is more strongly cespitose and has blue-gray foliage and ♀ scales about half as long as the perigynia. *Carex interior* has clearly beaked perigynia with swollen bases, and ♀ scales about half as long as the perigynia.

COMMENTS: *Carex tenuiflora* may form clumps or grow more loosely through moss mats, in the sun or in partial to full shade. It is rare in WA. Potential habitat in the northern tier of WA counties should be searched for this inconspicuous plant. The one known WA population lives in a sedge marsh and *Sphagnum* bog at the margins of a beaver pond. It is administratively protected. *Carex tenuiflora* is rare in most jurisdictions across the southern edge of its range, where its greatest threat is hydrologic change, including that caused by global warming. In some areas, logging, road building, and grazing by wild or domestic ungulates pose additional threats.

top left: pistillate scales,
   perigynia
top right: inflorescences
center left: habit
bottom: habitat

Carex tiogana D. Taylor and J. Mastrogiuseppe
Common name: Tioga Pass Sedge
Section: Chlorostachyae   Key: F

KEY FEATURES:
- Tiny sedge with falcate leaves and open inflorescence
- Dark, small perigynia with deciduous, white ♀ scales
- Rocky alpine benches with seepy gravel soils

DESCRIPTION: Habit: Cespitose. **Culms**: 1.8-15 cm tall. **Leaves**: 1.5-3 mm wide, falcate (arching), with serrulate margins and the midrib serrulate at least in the distal third. **Inflorescences**: Inflorescence bracts with well-developed sheaths. Lateral spikes ♀, with 3-9 perigynia. Lower spikes erect to nodding. Terminal spike ♂, 0.35-0.5 cm long. **♀ scales**: Mostly white with scabrous, gold midrib, obtuse to acuminate, shorter but wider than the perigynia, falling before the perigynia. **Perigynia**: Straw-colored to brown, oblong-ovate, veinless but with two marginal ribs, 1.2-2.1 mm long, 0.4-1 mm wide, usually with a few curved spines on the distal margins, with a beak 0.2-0.5 mm long. Stigmas 3. **Achenes**: Trigonous.

HABITAT AND DISTRIBUTION: Rocky benches in cirques, with thin soil kept wet by snowmelt, or coarse, wet, gravelly terraces by lake margins, near or above timberline, about 9000 feet elevation on Steens Mt., OR, and 9000 to 11000 feet elevation, usually on limestone, in Mono Co., CA.

IDENTIFICATION TIPS: *Carex tiogana* is a tiny, alpine sedge with arching leaves and small brown perigynia subtended by largely white ♀ scales that fall early. At first glance it might be dismissed as a small alpine grass, perhaps an *Agrostis*. *Carex tiogana* is the alpine relative of *C. capillaris*, a taller plant with flat leaves and larger perigynia. In *C. capillaris*, the midrib of the ♀ scales and the margins and midribs of its leaves are usually smooth.

COMMENTS: Rare alpine *C. tiogana* was discovered in the high Sierra Nevada Mts. in Mono County, CA. In 2001, similar dwarfed plants were found in two cirques at about 9000 feet on Steens Mt. in SE OR. Whether *C. tiogana* is best recognized as a species or as an intraspecific taxon within *C. capillaris* is an open question. Whatever taxonomic rank we may choose for it, this striking looking plant is isolated from typical *C. capillaris* and grows in an unusual habitat. Virtually nothing is known about the ecology of Oregon's tiny *C. tiogana* population, or what threats it may face.

Carex tiogana

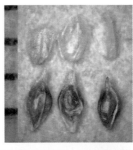

top left: perigynia, ♀ scales
top right: inflorescence
center left: habitat (alpine cirque)
bottom: habit

385

Carex tribuloides Wahlenb. var. *tribuloides*
Common name: Tribulation Sedge
Section: *Ovales*   Key: J

KEY FEATURES:
- Cespitose, with gynecandrous spikes and winged perigynia
- Perigynium wings abruptly narrowed or lacking below middle
- Leaf sheath fronts green and veined

DESCRIPTION: Habit: Cespitose. **Culms**: 50-110 cm tall, the vegetative culms sometimes rooting at the nodes and functioning as stolons. **Leaves**: 15-40 cm long, 3.2-7 mm wide. Leaf sheath fronts green and veined. **Inflorescences**: 2-5 (-8) cm long, 10-20 mm wide, brown, erect, often dense and head-like, sometimes elongated but dense distally. Spikes gynecandrous. ♀ **scales**: White-hyaline or silvery brown with green midstripe, (1.9-) 2.5-3 mm long, narrower than and half as long as the perigynia, tip acute to acuminate. **Perigynia**: Green to light brown, ovate-lanceolate to diamond-shaped, (3.3-) 3.6-5.4 mm long, 1.1-1.5 mm wide, with wing 0.1-0.3 mm wide, often asymmetric, abruptly narrowed and often lacking below the middle; with 3-6 dorsal veins and 2-4 ventral veins (which may be faint). Beak winged and ciliate-serrulate almost to the tip; 1.4-2 mm from achene top to beak tip. Stigmas 2. **Achenes**: Lenticular, 1-1.8 mm long, 0.6-0.9 mm wide, 0.3-0.5 mm thick.

HABITAT AND DISTRIBUTION: Floodplain forests, wet thickets, ditches, wet meadows; Rattlesnake Lake, WA, and lower Columbia River, OR. Native MN to NS, S to VA and KS, introduced to the PNW including the Fraser River in BC.

IDENTIFICATION TIPS: The aptly named *C. tribuloides* looks like many other species of section *Ovales*, until the green leaf sheath fronts are noticed. In the native *C. feta,* the most common species with green sheaths, the perigynia are ovate and mostly smaller, and the perigynium wings extend to the perigynium base. *Carex scoparia* has narrower leaves and usually longer perigynia.

COMMENTS: In the east, where it is more common and is sometimes a community dominant, *C. tribuloides* is used at a rate of 1-10% in seed mixes for wetland restorations, particularly for partially shaded areas.

Carex tribuloides var. *tribuloides*

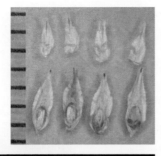

top left: pistillate scales, perigynia
top right, center left, and center right:
   inflorescences
bottom left: habitat (disturbed lake
   margin)

Carex tumulicola Mack.
Common name: Foothill Sedge
Section: *Phaestoglochin*   Key: H

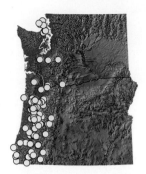

KEY FEATURES:
- Inflorescence coarse-textured, with brown and green scales
- Inflorescence usually bending to the side
- Forming patches in mesic or upland grasslands

DESCRIPTION: Habit: Rhizomatous, but the stout rhizomes short, so that the plant forms patches. **Culms**: 20-80 cm tall. **Leaves**: Green, tough, the widest 1.5-2.5 mm wide. **Inflorescences**: 1.5-5 cm long, 5-8 mm wide, usually bent somewhat to the side, with the lowest inflorescence bract usually longer than the lowest spike, sometimes longer than the inflorescence. Spikes androgynous, relatively narrow. ♀ **scales**: Tan or brown with green midrib, 3.3-5.2 mm long, about as long as the perigynium and somewhat wider, with a short (to 1 mm) awn. **Perigynia**: Light green or light brown, 3.5-5 mm long, 1.5-2 mm wide, the margins not flat (but with ribs), and with beak 1-3 mm long. Stigmas 2. **Achenes**: Lenticular.

HABITAT AND DISTRIBUTION: Grasslands, oak savanna, dry overgrazed slopes, and openings in forest, from low elevations to around 2400 feet, W of the crest of the Cascades and Sierra Nevada. S BC to S CA.

IDENTIFICATION TIPS: *Carex tumulicola* forms patches of tough green leaves in mesic to dry grasslands. At about knee height (well below the tops of most of the grasses) it produces brown inflorescences that usually bend to the side, often with the lowest bract diverging. Its perigynia are mostly hidden by the ♀ scales. Species in section *Ovales* have gynecandrous spikes with winged perigynia. *Carex hoodii* has a denser, relatively shorter inflorescence, and its perigynia are greenish with contrasting chestnut brown center (or light brown with contrasting dark brown center, when old). Sterile *C. tumulicola* might be confused with *C. rossii,* an upland species with reddish shoot bases. *Carex divulsa*, sold as *C. tumulicola* and naturalizing in the PNW, has longer inflorescences with the lower spikes separated.

COMMENTS: *Carex tumulicola* can be dominant in remnant native prairies. It is more resistant to trampling and grazing than are most PNW prairie species. On many hillsides that were overgrazed in the past it persists as the only native in a sea of annual weeds. It can be used for prairie restoration and erosion control. Both *C. tumulicola* and European *C. divulsa* are sold for garden borders and lawns under the names "Berkeley Sedge" and *C. tumulicola*. Both resist trampling, thrive when mowed, are evergreen, and survive with little watering, but only one is native to the PNW.

top left: pistillate scales, perigynia
top right: inflorescences
center left: habit
bottom: habitat

Carex unilateralis Mack.

Common name: One-sided Sedge
Section: *Ovales*   Key: J

KEY FEATURES:

- Cespitose, with gynecandrous spikes and winged perigynia
- Lowest inflorescence bract usually long, leaf-like and erect
- Inflorescence dense, tilted to the side
- Marshes, ditches, seasonally wet meadows

DESCRIPTION: Habit: Cespitose. **Culms**: 35-75 cm tall. **Leaves**: 2-3 (-4) mm wide. **Inflorescences**: Inflorescence bracts green, leaf-like, longer than the inflorescence, which is dense and head-like, usually tilted to one side, (1-) 1.4-2.5 cm long, 9-17 mm wide. Spikes gynecandrous. ♀ **scales**: Gold to brown with midrib green or gold, 3.3-4.8 mm long, shorter and narrower than the perigynia. **Perigynia**: Cream-colored to light brown, ovate to lanceolate, 3.5-5 mm long, 1.3-1.75 mm wide, 0.3-0.5 mm thick, with 0-11 dorsal veins, with wing 0.2-0.3 mm wide. Beak tip usually winged and ciliate-serrulate to the tip, but occasionally unwinged, brown, parallel-sided, and entire for the distal 0.5-0.6 mm; (1.4-) 1.7-2.5 mm from top of achene to tip of beak. Stigmas 2. **Achenes**: Lenticular, (1.2-) 1.5-1.9 mm long, 0.75-1 (-1.2) mm wide, 0.3-0.4 mm thick.

HABITAT AND DISTRIBUTION: Wet prairie, vernal pools, pond margins, marshes, road ditches, and other seasonally wet places, W of the crest of the Cascades but rare near the coast, 0 to 3000 feet. S BC to CA, introduced to Japan.

IDENTIFICATION TIPS: With its long bracts and inflorescence angled to one side of the culm, *C. unilateralis* is usually easy to identify. Some plants are intermediate between this and *C. athrostachya,* which lives mainly E of the Cascades but occasionally on the W side. *Carex athrostachya* inflorescences are erect, perched directly atop the culm, and its lower inflorescence bracts are more bristle-like and speading. Its beaks are unwinged, brown, and entire, but in *C. unilateralis* the majority of beaks are flattened and ciliate-serrulate almost to the tip.

COMMENTS: *Carex unilateralis* can be codominant in westside wet prairie with *C. densa,* Tufted Hairgrass (*Deschampsia cespitosa*), and Meadow Barley (*Hordeum brachyantherum*). *Carex unilateralis* is used for revegetating wet prairie and the margins of shallow or seasonal wetlands W of the Cascades. It survives fire well but above-ground biomass and seed production decrease for the first year or two after the fire. *Carex unilateralis* is a confirmed host of the Dun Skipper, a butterfly which oviposits only on *Carex.*

top left: habit
top middle: pistillate
   scales, perigynia
top right and center left:
   inflorescences
bottom: habitat (wet
   prairie)

Carex utriculata Boott

Common name: Southern Beaked Sedge
Section: *Vesicariae*   Key: E
Synonyms: *C. rostrata* L., misapplied

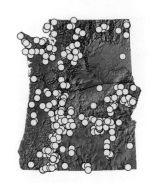

KEY FEATURES:
• Leaf sheath with crosswalls between veins
• Rhizomatous stands growing in water
• Leaves green, not papillose
• Spikes with "corncob" appearance

DESCRIPTION: Habit: Rhizomatous, forming large stands. **Culms**: 25-120 cm tall. **Leaves**: Green, flat to broadly W-shaped in cross section, the widest leaves 4.5-12 (-15) mm wide, with the surface smooth, not papillose. Leaf sheaths often brown to tinged reddish, somewhat spongy, with crosswalls common between the vertical veins, making a pattern like brickwork. **Inflorescences**: Lateral 2-5 spikes ♀, erect to ascending, mostly 2-10 cm long, with 8 spiraling columns of perigynia; well separated from the 2-5 terminal ♂ spikes, which are mostly 2-7 cm long. ♀ **scales**: Mostly shorter than the perigynia, apex acute to acuminate, awnless. **Perigynia**: Spreading, inflated, green, straw-colored, or reddish brown, with 9-15 veins, (3.2-) 4-8.6 mm long, 1.7-3 mm wide, the apex abruptly narrowed to a beak (1-) 1.2-2.7 mm long, with straight teeth 0.2-1.3 mm long. Stigmas 3, styles persistent, becoming curved. **Achenes**: Trigonous.

HABITAT AND DISTRIBUTION: Lake shores, marshes, streamsides, bogs, fens, with high water table, usually on organic soils; throughout OR and WA. AK to NF, S to CA, MN, TN, and VA, also Mexico and eastern Eurasia; absent from the Great Plains.

IDENTIFICATION TIPS: *Carex utriculata* is a common, robust species forming dense stands in shallow water. It has large, inflated perigynia with short, straight beak teeth. The backs and sides of its leaf sheaths have many crosswalls between the vertical veins. *Carex vesicaria* and *C. exsiccata* are similar but they lack this "brickwork" of crosswalls and have perigynia that taper less abruptly to the beak. *Carex atherodes* has hairs on the leaves and leaf sheath fronts, and long, divergent beak teeth. In NE WA, see *C. rostrata* and watch for *C. lacustris,* with ligules longer than wide and perigynia with a gradually tapered beak 0.5-1.6 mm long.

COMMENTS: Most literature on *C. utriculata* ecology uses the misapplied name *C. rostrata*. *Carex utriculata* can tolerate flooding up to 16 inches in spring, groundwater to 2 feet below the surface in later summer, heavy metal contamination, and acidic soils. Productivity is high and most ungulates graze it in spring; the tough mature leaves are less palatable. The rhizomes have been used for basketry.

Carex utriculata

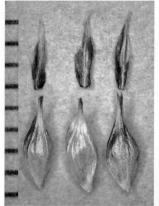

top left: pistillate scales, perigynia
top right: inflorescence
center left: leaf sheaths (brick pattern)
bottom: habit, habitat

Carex vallicola Dewey
Common name: Valley Sedge
Section: *Phaestoglochin*   Key: H

KEY FEATURES:
- Shiny, plump perigynia with marginal ribs displaced to the ventral side
- Short, few-flowered spikes
- Dry grasslands E of the Cascades

DESCRIPTION: Habit: Cespitose or very short-rhizomatous. **Culms**: 12-60 cm tall, light brown or warm tan at base. **Leaves**: 1-3 mm wide, inconspicuously papillose ventrally. **Inflorescences**: 0.5-3 cm long, 4-8 mm wide, with 5-10 short, crowded, androgynous spikes. ♀ **scales**: Whitish or pale brown hyaline with green center, the body slightly shorter than or as long as the perigynia, with tip acute or short-awned. **Perigynia**: Light brown, shiny, oval, broadly rounded on the dorsal side so that the two marginal ribs appear to be on the flat ventral surface, with several (7-15) more or less obscure dorsal veins and no ventral veins, 3.3-4 mm long, 1.8-2.3 mm wide, with a more or less spongy area at base that is 0.7-1 mm long and dries wrinkled; perigynium abruptly contracted to a short (0.5-1 mm long) beak that is smooth-margined or serrulate. Stigmas 2. **Achenes**: Lenticular.

HABITAT AND DISTRIBUTION: Mesic depressions and hill slopes in dry grassland or sage steppe, generally in gravelly loam, often with Big Sage (*Artemisia tridentata*), 4300-7200 feet elevation, Okanogan Co., WA, E OR; one old NW OR record introduced along a railroad track. BC to SD, S to CA, NM and Mexico.

IDENTIFICATION TIPS: *Carex vallicola* is a cespitose sedge with a fairly narrow inflorescence composed of short spikes. Its shiny, plump perigynia have the marginal veins displaced to the ventral side. It often grows with *C. hoodii,* which has darker, more compact inflorescences, and perigynia that have marginal veins at the margins where they belong and are green with copper centers (light brown with dark brown centers when fully mature). Sterile plants might be confused with *C. rossii,* which has reddish shoot bases.

COMMENTS: *Carex vallicola* is palatable to livestock but declines when too heavily grazed and is generally absent from overgrazed ranges. Populations are small. The few WA populations are vulnerable to recreational impacts, especially by off-road vehicles.

top left: perigynia
top right: inflorescences
center left: habit
bottom: habitat

Carex vernacula L. H. Bailey

Common name: Native Sedge
Section: (controversial)   Key: H
Synonyms: *C. foetida* Allioni var. *vernacula* (L. H. Bailey)
  Kükenthal

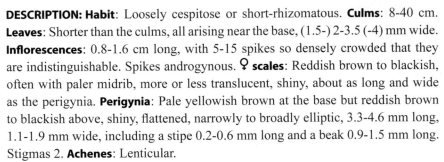

KEY FEATURES:
• Inflorescence a dense, dark, truncate-based ball, the
  spikes indistinguishable
• Perigynia stipitate
• Alpine and subalpine wetlands

DESCRIPTION: Habit: Loosely cespitose or short-rhizomatous. **Culms**: 8-40 cm.
Leaves: Shorter than the culms, all arising near the base, (1.5-) 2-3.5 (-4) mm wide.
Inflorescences: 0.8-1.6 cm long, with 5-15 spikes so densely crowded that they
are indistinguishable. Spikes androgynous. ♀ **scales**: Reddish brown to blackish,
often with paler midrib, more or less translucent, shiny, about as long and wide
as the perigynia. **Perigynia**: Pale yellowish brown at the base but reddish brown
to blackish above, shiny, flattened, narrowly to broadly elliptic, 3.3-4.6 mm long,
1.1-1.9 mm wide, including a stipe 0.2-0.6 mm long and a beak 0.9-1.5 mm long.
Stigmas 2. **Achenes**: Lenticular.

HABITAT AND DISTRIBUTION: Alpine and subalpine wet meadows, rocky slopes
that receive snowmelt, edges of headwater streams, and lakeshores; Mt. Adams,
WA, Mt. Hood, Wallowa Mts., Steens Mt., and Lake Co., OR. WA to WY, S to
CA and CO.

IDENTIFICATION TIPS: This is an alpine sedge with dense, dark heads, dark,
stipitate perigynia, and long, persistent anthers. The *Ovales* are similar but have
gynecandrous spikes and winged perigynia. *Carex illota* has smaller inflorescences
and uniformly dull dark brown perigynia with no stipes. *Carex jonesii* has the
perigynium base swollen with pithy tissue.

COMMENTS: *Carex vernacula* can be a community dominant along small alpine
streams. It has been recommended for roadside erosion control plantings in
appropriate habitats in NV and CA, where it is more common than in the PNW.
Its small stature and dense, dark heads make it an attractive addition to a moist
rock garden or margin of a garden pool. It is grazed readily by domestic sheep.
Carex vernacula is very similar to European *C. foetida* and is often considered
a variety of that species. Recent molecular phylogenetic work suggests that it
is more closely related to *C. jonesii,* which is common in wetland habitats at
moderate to high elevations in mountains of western N America.

top left: perigynia
top right: inflorescences
center right: habit, habitat (alpine
  fellfield)
bottom: habit (very old plant)

Carex vesicaria L.

Common name: Inflated Sedge
Section: *Vesicariae*   Key: E
Synonyms: *C. vesicaria* L. var. *vesicaria, C. monile*
   Tuckerm.

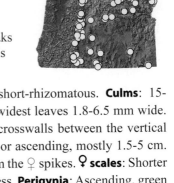

KEY FEATURES:
• Perigynia inflated, contracting to the distinct beaks
• Leaf sheaths without a "brickwork" of crosswalls
• Cespitose

DESCRIPTION: Habit: Cespitose, or sometimes short-rhizomatous. **Culms**: 15-100 cm tall. **Leaves**: Green, V- or W-shaped, the widest leaves 1.8-6.5 mm wide. Basal leaf sheath reddish, not spongy, with few crosswalls between the vertical veins. **Inflorescences**: Lateral 1-3 spikes ♀, erect or ascending, mostly 1.5-5 cm. Terminal 1-3 spikes ♂, 2-7 cm, well separated from the ♀ spikes. ♀ **scales**: Shorter than the perigynia, apex acute to acuminate, awnless. **Perigynia**: Ascending, green to straw-colored or reddish brown, with 7-12 veins, not leathery, (3.5-) 4-7.5 (-8.2) mm long, (1.4-) 1.7-3.3 mm wide, contracted to the distinct beak 1.1-3 mm long, with straight teeth 0.3-0.7 mm long. Stigmas 3, styles persistent, becoming curved. **Achenes**: Trigonous.

HABITAT AND DISTRIBUTION: Pond margins, lake shores, wet spots in meadows, generally in shallow water, (100-) 3000-9000 ft elevation; mainly E of the crest of the Cascades in OR and WA. BC to MT, S to CA and WY, also MB to NF, S to MO and VA, also interruptedly circumboreal.

IDENTIFICATION TIPS: *Carex vesicaria* is a large, cespitose species with relatively large, somewhat inflated perigynia that have distinct beaks and short, straight teeth. It is similar to *C. exsiccata,* which grows W of the Cascades and has longer perigynia that taper more gradually to the tip. Immature plants can be identified only by geography. Some plants along the crest of the Cascades are intermediate. *Carex utriculata* differs because it is rhizomatous and has leaf sheaths with a "brickwork" of crosswalls. Where the two grow together, *C. utriculata* grows in deeper water.

COMMENTS: *Carex vesicaria* is often a community dominant in shallow water of lakes and ponds. It is more palatable to livestock than coarser *C. utriculata.* Nutrient release after fire can increase growth, but organic soils can burn, killing rhizomes. Major threats are draining and filling of wetlands. Lowered water tables favor competing Reed Canarygrass (*Phalaris arundinacea*). *Carex vesicaria* is used for habitat restoration of lakeshores, wet meadows, and backwaters of the Columbia River. Variation in *C. vesicaria* has led to contradictory taxonomic treatments; a worldwide taxonomic review is needed.

top left: pistillate scales, perigynia, achenes
top right: inflorescences
bottom: habit, habitat

Carex viridula Michx. ssp. *viridula*
Common name: Green Sedge
Section: *Ceratocystis*   Key: F
Synonyms: *C. oederi* Retz.

KEY FEATURES:
- Perigynia strongly spreading, green
- Spikes crowded
- Lowest inflorescence bract widely spreading

DESCRIPTION: Habit: Cespitose. **Culms**: 2-40 cm tall. **Leaves**: 1-3.1 (-4.5) mm wide. **Inflorescences**: Lowest inflorescence bract diverging at a right angle or more from the culm. This bract usually lacks a sheath, but if the lowest lateral spike is remote from the others, its bract may have a sheath > 4 mm long. Lateral spikes 1-8, ♀ or androgynous, 0.4-1.4 cm long, 3-11 mm wide, crowded or sometimes the lowest remote. Terminal spike ♂ or androgynous, sessile or pedunculate, 0.8-2.5 cm long, 1-3.3 mm wide. ♀ **scales**: Shorter than perigynia. **Perigynia**: Widely spreading, yellowish green to green, (1.8-) 2.2-3.3 (-3.9) mm long, 0.8-1.6 mm wide, with beak 0.3-1.3 mm long, straight or slightly bent toward the dorsal side. Stigmas 3. **Achenes**: Trigonous.

HABITAT AND DISTRIBUTION: Several distinct habitats in the PNW: deflation plains among coastal sand dunes, margins of hot springs in SE OR and ID, bogs in mountains of N WA, a serpentine fen in CA; elsewhere, populations grow in other habitats, including diverse fens. AK to NF and Greenland, S to CA, NM, IL, and NJ, also Europe, Morocco, mts of central Asia, Japan, and Kamchatka.

IDENTIFICATION TIPS: *Carex viridula* is a strictly cespitose sedge with crowded green spikes, spreading perigynia, and a divergent inflorescence bract. *Carex flava* has longer perigynia that are strongly reflexed, with beaks curved back toward the dorsal side.

COMMENTS: *Carex viridula* is part of a widely distributed group of more or less distinct populations that can be treated as poorly differentiated species or as subtaxa within *C. viridula*. North American plants show little genetic variation compared to the highly variable European populations. Probably this species arose in Europe and invaded N America relatively recently. Many extant populations are isolated, apparently left behind as the species retreats northward with the glaciers. The small coastal plants are cute, growing as dense tufts with clusters of green spikes. They would make an interesting addition to a wet garden, although they might become leggy if fertilized. Inland plants tend to be taller with less compact inflorescences.

Carex viridula ssp. *viridula*

top left:
  perigynia
*top right and
  center left*:
  inflorescences
bottom: habit,
  habitat

Carex vulpinoidea Michx.
Common name: Fox Sedge
Section: *Multiflorae*   Key: H

KEY FEATURES:
- Cross-corrugated leaf sheath fronts
- Long, interrupted inflorescence with long, hair-like bracts
- Mainly E of Cascades

DESCRIPTION: Habit: Cespitose. **Culms**: 30-100 cm tall. **Leaves**: 5 mm wide, generally longer than the flowering culms. Leaf sheath fronts cross-corrugated, white-hyaline with brown spotting. **Inflorescences**: Elongate, (3-) 7-10 cm long, 15 mm wide, somewhat interrupted, with hair-like inflorescence bracts conspicuous throughout. Spikes many, short, androgynous. Lowest node usually with a short branch with at least 1 side spike, thus the lowest node often seeming to produce 2 spikes. ♀ **scales**: Brown, short-awned. **Perigynia**: Dull grayish brown, the body elliptic to ovate, 2-3.2 mm long, 1.3-1.8 mm wide, with 0-3 dorsal veins, veinless ventrally. Base obtuse or rounded, depending on the amount of pithy tissue within. Stigmas 2. **Achenes**: Lenticular.

HABITAT AND DISTRIBUTION: Moist to wet meadows, marshes, and ditches; mainly E of the Cascades in WA and OR, but also growing in the lower reaches of the Columbia River, probably introduced elsewhere W of the Cascades. S BC to NF, S to CA, FL, and Mexico; introduced to CA, Europe, Japan, and New Zealand.

IDENTIFICATION TIPS: *Carex vulpinoidea* is a cespitose marsh plant with cross-corrugated leaf sheaths and long, interrupted inflorescences with conspicuous though almost hair-thin bracts. It is most similar to *C. densa,* its equivalent W of the Cascades. *C. densa* has a shorter, uninterrupted inflorescence with less conspicuous bracts, and its yellower perigynia average larger (2.8-4 mm long) with more dorsal veins and ventral veins. Its flowering culms are longer than the leaves. *Carex vulpinoidea* is superficially similar to *Ovales* sedges, which have gynecandrous spikes, but easily distinguished by the cross-corrugated leaf sheath fronts.

COMMENTS: Throughout its range, this common wetland species is used in wetland restoration projects, rain gardens, and stormwater retention ponds because it holds soil well and provides food for waterfowl. It is well suited for wetland restorations E of the Cascades, but using it W of the Cascades creates the potential for gene flow from *C. vulpinoidea* into closely related *C. densa,* which is native on the W side. Populations along the lower Columbia River may have originated from seed washed down the river from eastside populations.

top left: perigynia *top right*: inflorescences
center: cross-corrugated leaf sheath front *bottom:* habit, habitat

Carex whitneyi Olney

Common name: Whitney's Sedge
Section: *Longicaules*   Key: F
Synonyms: *C. jepsonii* J.T. Howell

KEY FEATURES:
- Hairy leaves
- Elliptic, veined, glabrous perigynia
- Openings in dry, upland, conifer forest

DESCRIPTION: Habit: Cespitose. **Culms**: 25-100 cm tall, much longer than the leaves. **Leaves**: 2-6 mm wide, pubescent. **Inflorescences**: Lateral spikes 2-3, ♀, 0.7-3 cm long, stalked, with leaf-like bracts. Terminal spike ♂, 0.5-3 cm long. ♀ **scales**: White-hyaline with green midstripe, shorter than the perigynia, usually with a short awn. **Perigynia**: Green, ovate, veined, glabrous, 3.7-5.5 mm long, with beak 0.2-1 mm long and teeth less than 0.5 mm long. Stigmas 3. **Achenes**: Trigonous.

HABITAT AND DISTRIBUTION: Dry meadows and openings in conifer forest, on sandy to gravelly soils, 5000-12,000 feet elevation, mts of Jackson, Lake, Klamath Cos., S OR, and one record each in Deschutes Co. and on Mt Hood, OR. Also CA, NV.

IDENTIFICATION TIPS: Its hairy leaf blades make *C. whitneyi* easy to identify. The few other PNW sedges with hairy leaf blades live in moist habitats. They are *C. pallescens* and *C. hirsutella*, with almost beakless, weakly veined perigynia, and *C. gynodynama,* a low-elevation species with sparse spreading hairs on the perigynia. *Carex mendocinensis* leaves may have a few hairs.

COMMENTS: *Carex whitneyi* is an early-successional species that colonizes disturbed sites in conifer forest. The Mt Hood and Deschutes Co. populations, both discovered since 1990, may be native there or introduced by hikers. In order to determine its status, *C. whitneyi* should be sought elsewhere in the Cascades. This unusual sedge is one of only four in section *Longicaules,* all with hairy leaf blades. The other three all grow in Mexico. *Carex whitneyi* was named after Josiah Dwight Whitney, state geologist of California 1860-1876, after whom Mt. Whitney was named. *Carex whitneyi* would make an attractive garden subject.

top left: perigynia
top right: inflorescences
center right: hairy leaf
bottom: habit

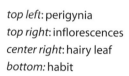

Kobresia myosuroides (Villars) Fiori

Common name: Mouse-tail Kobresia
Key: A, D
Synonyms: *Carex myosuroides* Villars, *Kobresia bellardii*
(Allioni) Degland ex Loiseleur

KEY FEATURES:
• Perigynia open on one side
• Inflorescence simple, like a single long spike
• Very narrow leaves
• Dry to mesic alpine slopes and meadows

DESCRIPTION: Habit: Densely cespitose to mat-forming. **Culms**: 5-20 (-35) cm long, little longer than the leaves. Dead, shiny, brown leaf sheaths abundant at the base of the plant. **Leaves**: 2-20 cm long and very thin, 0.2-0.5 mm wide. **Inflorescences**: Narrow and cylindrical, 1-3 cm long, 2-3 mm wide, unbranched. Lower perigynia bisexual, upper ones ♂. **Scales**: Brown with hyaline margins, 2-3.5 mm long, the midvein distinct almost to the tip, which is obtuse or minutely pointed. **Perigynia**: Brown, 2-3.5 mm long, open to the base, each containing a ♀ flower (producing the achene) and above it a small ♂ flower, or the upper ones with just 1 ♂ flower. Stigmas 3. **Achenes**: Trigonous.

HABITAT AND DISTRIBUTION: Dry to mesic (occasionally moist) arctic, alpine, and subalpine slopes and meadows, on diverse substrates, often on windswept slopes with little snow accumulation; Wallowa Mts. and Steens Mt., OR. AK to NF, S to CA and AZ.

IDENTIFICATION TIPS: The unusual, narrow, cylindric inflorescences of *Kobresia myosuroides* might suggest a male *C. scirpoidea,* but that plant has hairy scales and wider, flat leaves. The very narrow leaves and brown, persistent leaf sheath of *K. myosuroides* are difficult to distinguish from sterile *C. nardina* or *C. filifolia,* which may grow in the same habitat. *Kobresia myosuroides* may mingle with *C. nigricans* and *C. subnigricans*. Those two *Carex* are rhizomatous and each culm has a thicker spike. In addition, *C. nigricans* has wider leaves.

COMMENTS: In the cold alpine meadows inhabited by *K. myosuroides*, decomposition is so slow that nitrogen availability often limits plant growth, but *K. myosuroides* absorbs nitrogen in the form of amino acids and via ectomycorrhizal fungi. Individual shoots often live for 5 years. During the first 3 years, each shoot produces new leaves. In the 4th year, it produces daughter shoots, and in the 5th year, it flowers, sets seed, and dies. Estimated age of the older *K. myosuroides* clumps vary from 70 to 250 years. Alpine pikas graze *K. myosuroides* heavily.

top left: pistillate scales, perigynia
top right: inflorescences
center left: habit
bottom: habitat (alpine cirque)

Kobresia simpliciuscula (Wahlenb.) Mack.
Common name: Simple Kobresia
Key: D
Synonyms: *Carex simpliciuscula* Wahl.

KEY FEATURES:
- Perigynia open on one side
- Inflorescence consisting of distinct spikes
- Leaves narrow but not thread-like
- Alpine bogs and marshes

DESCRIPTION: Habit: Densely cespitose to mat-forming. **Culms**: 5-35 cm long, longer than the leaves. **Leaves**: 2-20 cm long and 0.2-1.5 (-2) mm wide, green. **Inflorescences**: (8-) 1-3.5 cm long, (2-) 3-8 mm wide, branched, with short spikes. In each spike, lower perigynia bisexual and upper ones ♂. **Scales**: Brown with hyaline margins, 2-3 mm long, the midvein distinct almost to the tip, which is obtuse or acute. **Perigynia**: Brown, 2.5-3.2 mm long, open to the base, each containing a ♀ flower (producing the achene) and above it a small ♂ flower, or the upper ones with just 1 ♂ flower. Stigmas 3. **Achenes**: Trigonous.

HABITAT AND DISTRIBUTION: Bogs and fens, generally on calcareous substrates, near and above timberline; NE OR. AK to NF, S in the Rockies to UT and CO. Circumboreal but with a patchy distribution and rare in much of its range.

IDENTIFICATION TIPS: *Kobresia simpliciuscula* looks very much like a young *Carex* until it becomes clear that the cellophane-textured bract surrounding the achene is open on one side. *Kobresia myosuroides* differs in having a single, narrow, cylindrical, spike-like inflorescence. *Kobresia myosuroides* grows in drier sites. *Carex nardina* has similar very narrow leaves and build-up of dead, brown sheaths at bases of shoots, but has closed perigynia and grows on drier sites.

COMMENTS: This circumboreal species probably once had a wide range at our latitudes but has retreated N and up slopes as the glaciers have retreated. It is now rare in much of its range. Other than global warming, there are few and probably only minor threats to the PNW populations. Hydrologic alteration of its wet habitat by mining, logging, road building, and recreational activities threaten some populations in other regions. *Kobresia simpliciuscula* appears to have resistance to moderate grazing, but associated trampling and soil compaction can impact it negatively. It is adapted to the low-nutrient regime characteristic of botanically diverse bogs and fens. Addition of plant nutrients favors faster-growing grass species and can result in competitive exclusion of *K. simpliciuscula*.

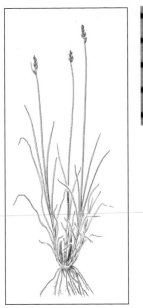

top left: habit
top middle: scales, perigynia
top right: inflorescence
bottom: habit

Sedges to Watch for in Oregon and Washington

Carex buchananii **Bergg.** (Leatherleaf Sedge, Bronze Sedge). Densely cespitose. Leaves bronze or orange-brown (olive green in shade), wiry, 1-1.5 mm wide, often curling at the tips. Lateral spikes pistillate (rarely gynecandrous), 0.5-3 cm, terminal 1(-2) spikes staminate. Pistillate scales whitish, shorter than or as long as the darker perigynia, which are lenticular, 2.5-3 mm long and faintly nerved. Stigmas 2, achenes lenticular. Cultivated and escaping; native to New Zealand. Two other New Zealand *Carex* commonly cultivated here have bronze foliage. *Carex testacea* has more strongly nerved perigynia and wider leaves 1-2.5(-3) mm wide. The bronze form of *C. comans* has trigonous achenes.

Carex comans **Berggr**. (New Zealand Hair Sedge). Densely cespitose, forming a dense mop of very narrow, drooping, leaves that are 0.5-1(-1.5) mm wide and can be green, gray-green, blue-green, bronze, yellowish, or brown. Lateral 5-7 spikes pistillate, 0.5-2.5 cm long; terminal 1-2 spikes staminate. Pistillate scales usually slightly shorter than the perigynia, brown to reddish brown. Perigynia 2.5-3.5 mm long, smooth when fully mature. Stigmas 3, achenes trigonous. Cultivated and escaping, native to New Zealand.

***Carex nigra* (L.) Reichard** (Smooth Black Sedge). Rhizomatous, forming clumps, patches, or extensive stands. Leaves 2-4.5 mm wide, papillose on upper surface. Lowest inflorescence bract subequal to the inflorescence. Lower 2-3 spikes pistillate, 1.2-4.2 cm; upper 1-2 spikes staminate. Pistillate scales dark purple-brown to black, < perigynia. Perigynia strongly papillose, 2-3.7 mm, green to pale brown with red-brown or black spots, with 3-9 veins on each face, often bristly on the very tip. Stigmas 2, achenes lenticular. Habitat wet meadows, ditches, and bogs in SW BC, likely to spread to NW WA; native to Eurasia. *Carex aquatilis* and *C. kelloggii* typically have longer lowest inflorescence bracts. Also, *C.*  *kelloggii* is cespitose. *Carex aquatilis* lacks veins on the perigynium faces; it hybridizes with *C. nigra*.

***Carex pumila* Thun.** (Dwarf Sedge). Strongly rhizomatous. Culms short (8–30 cm), often buried in sand up to the inflorescence. Leaves 1.5-3.5 mm wide. Terminal 1-3 spikes staminate; lateral 1-4 spikes pistillate, short-cylindric, 1.1-4.5 cm long, all overlapping and ascending. Pistillate scales acute to awned, with scabrous margins but otherwise glabrous. Perigynia  veined, thick-walled and corky, ovoid, 4.5-8 mm long, with beak 0.8-1.6 mm long, style persistent. Stigmas 3, achenes trigonous. An inconspicuous species of disturbed, sandy soils, apparently arrived on ship ballast in Portland, where it was collected from 1906 to 1916 but not documented since.

Sedges with Distinctive Traits or Habitats

Your sedge may have distinctive characteristics and/or habitat that limit the number of possible species you need to consider for identification. PNW sedges with distinctive traits or habitats are listed here. Don't forget to confirm the plant's identity by keying it. Some very rare introduced species or species that rarely exhibit the unusual trait are not included in the lists. Material in parentheses gives range or other information that can further limit the possibilities.

Leaves hairy
C. atherodes (E)
C. hirsutella (W OR)
C. hirta (W)
C. gynodynama (SW OR)
C. mendocinensis (SW OR)
C. pallescens
C. sheldonii (E)
C. whitneyi (Cascades, SW OR)

Leaves very broad, more than 1 cm wide
C. amplifolia
C. hendersonii (W)
C. lacustris (NE WA)
C. pendula (W)
C. utriculata
Scirpus microcarpus (not included in this book)

Leaf sheaths cross-corrugated
C. cusickii (occasionally)
C. densa (W)
C. jonesii (rarely)
C. neurophora
C. stipata var. *stipata*
C. vulpinoidea

Culms gigantic, sometimes over 5 feet long
C. atherodes
C. cusickii
C. obnupta
C. pendula

Perigynia gigantic
C. kobomugi (rhizomatous, tough; sand)
C. macrocephala (rhizomatous, tough; sand)

Stoloniferous, perhaps only late in season
C. chordorrhiza
C. feta
C. limosa
C. longii
C. leporina
C. scoparia var. *scoparia*
C. tribuloides

Spikes more than 6 cm long
C. amplifolia
C. aquatilis var. *dives*
C. obnupta
C. pendula

Perigynia hairy; habitat uplands
C. brainerdii (with basal spikes; SW OR)
C. brevicaulis (with basal spikes; coastal)
C. concinna (NE OR)
C. concinnoides (4 stigmas)
C. deflexa var. *boottii* (with basal spikes)
C. filifolia var. *filifolia* (single spike)
C. halliana (Cascades)
C. hirta (hairy leaves; Portland)
C. inops ssp. *inops*
C. rossii (with basal spikes)
C. scabriuscula (SW OR; single spike)
C. scirpoidea
C. serpenticola (SW OR)

Uplands in sagebrush steppe
C. douglasii
C. filifolia var. *filifolia* (single spike)
C. petasata
C. rossii (short; hairy perigynia)
C. vallicola

Perigynia hairy; habitat wetlands

C. atherodes (hairy leaves; E)
C. concinna (NE OR)
C. gynodynama (hairy leaves; SW OR)
C. hirta (hairy leaves; Portland)
C. lasiocarpa
C. pellita
C. sheldonii (hairy leaves; E)

Perigynia very hard, almost beakless

C. lyngbyei (coastal salt marsh)
C. obnupta (freshwater; W)

Stigmas 4

C. concinnoides
C. scabriuscula (occasionally)

Coastal sand dunes and deflation plains

C. brevicaulis (dry dunes)
C. kelloggii var. *limnophila* (wet places)
C. kobomugi (dry dunes)
C. macrocephala (dry dunes)
C. obnupta (wet spots)
C. pansa (dry dunes)

Coastal salt marsh dominant in extensive single-species stands

C. lyngbyei

Last sedge standing in heavily grazed sites

C. nebrascensis (E)
C. obnupta (W)
C. praegracilis (E)

Scour zone of large rivers

C. interrupta (rhizomatous, in sand)
C. nudata (dense tussocks in rocks)

Serpentine fens (often with *Darlingtonia*)

C. echinata ssp. *echinata*
C. hassei
C. klamathensis
C. mendocinensis
C. serratodens

Dry conifer forest (in openings)

C. brainerdii (SW OR)
C. concinnoides
C. deflexa var. *boottii*
C. geyeri
C. inops ssp. *inops* (Cascades & W)
C. multicaulis (SW OR)
C. rossii
C. serpenticola (serpentine; SW OR)
C. siccata (E WA)
C. whitneyi (S Cascades)

Moist forest (soft, loosely cespitose plants)

C. bolanderi
C. brunnescens (WA)
C. deweyana var. *deweyana* (N WA)
C. hendersonii (loose sheaths)
C. infirminervia
C. laeviculmis
C. leptopoda

Moist alkaline sites, east of Cascades

C. atherodes (leaves and perigynia often hairy)
C. douglasii (dioecious)
C. nebrascensis
C. praegracilis (dioecious)
C. sheldonii (hairy leaves and perigynia)
C. simulata

Perigynia yellow to orange, succulent, pumpkin-like

C. aurea

Ovales of lowlands east of Cascades

C. athrostachya
C. bebbi
C. brevior (WA)
C. microptera
C. pachystachya
C. petasata
C. praticola
C. scoparia (WA)
C. subfusca
C. sycnocephala

Ovales of non-coastal lowlands west of Cascades

C. athrostachya
C. feta
C. leporina
C. pachystachya
C. praticola
C. scoparia var. *scoparia*
C. subfusca
C. unilateralis

Ovales of the coast

C. crawfordii
C. harfordii
C. longii
C. pachystachya
C. subbracteata
C. unilateralis

Ovales of uplands in the alpine zone

C. haydeniana
C. microptera
C. phaeocephala
C. proposita
C. straminiformis
C. tahoensis

Ovales of dry uplands (not alpine)

C. davyi
C. integra
C. microptera
C. pachycarpa
C. petasata (shrub steppe)
C. preslii
C. straminiformis

Glossary

Abaxial: dorsal.

Acuminate: with the tip with concave sides, tapering to a narrow point.

Acute: with the tip formed by the straight or convex sides coming together at an angle smaller than 90°.

Adaxial: ventral.

Allopatric: (of two or more species) living in different areas, not together.

Alpine: above tree line on mountains.

Androgynous: with male flowers above female flowers in the same spike. (Compare *gynecandrous*)

Aphyllopodic: with lower leaves of the fertile culm reduced to bladeless sheaths. (Compare *phyllopodic*)

Appressed: (of perigynia) oriented not really parallel to the axis of the spike, but less than about 45° from it.

Ascending: with the tips pointing upward; (of perigynia) oriented about 45° from the axis of the spike. (Compare *spreading*)

Axil: the "armpit," the point of attachment of, for example, a bract and the culm. A perigynium arises in the axil of the pistillate scale.

Basal spike: a spike whose peduncle arises at the base of the culm.

Beak: the narrowed part of the perigynium above the body, often with 2 teeth at the tip.

Bidentate: with 2 teeth, as in the beaks of some perigynia.

Blade: the usually flat part of the leaf that extends beyond the leaf sheath.

Body: the main part of an object, usually the main part of the perigynium, excluding the beak and stipe (if any).

Bract: a modified, often reduced leaf. (See *scale*)

Bristle-like: (of an inflorescence bract) very thin and not leaf-like, consisting of little more than the midrib. This does not imply that it is hard or sharp.

Cespitose: with a bunchgrass-type growth form, all the culms arising together in a tuft, rhizomes seemingly absent. Some species of cespitose sedges may be densely tufted, and others more loosely clustered.

Channeled: (of a leaf) with the midrib sunken in a deep longitudinal groove.

Ciliate: with hairs (cilia) on the edge.

Cortex: the part of the root between its epidermis and the strand of vascular tissue in the center.

Cross-corrugated: (of a leaf sheath) wrinkled, with alternating ridges and furrows, (= cross-rugulose, = cross-wrinkled).

Culm: the stem-like part of a grass, sedge, or rush; can be fertile or sterile.

Deciduous: falling off. (Compare *persistent*)

Deciduous style: When young, a deciduous style is marked by an abrupt change of color where it will later fall, leaving a small stub at the top of the achene. (Compare *persistent style.*)

Disjunct: separated. A plant's range is disjunct if some of the populations are located very far from others.

Dioecious: with male and female flowers on different plants.

Distal: at the end farther away from the base of the plant, at the upper end, toward the tip. (Compare *proximal*)

Dorsal: the surface that is farther away from the stem, the back surface. (Compare *ventral*)

Elliptic: oval-shaped, widest near the middle.

Entire: smooth-edged. (Compare *serrulate*)

Enveloping: wrapping around (e.g., perigynium tightly enveloping achene).

Erose: appearing eroded, worn, or torn.

Exserted: sticking out from, projecting from.

Fibrillose: disintegrating and leaving the veins intact. (See *ladder-fibrillose*)

Filament: the thin, pale stalk that attaches the anther to the flower. When the anthers have fallen, the filaments usually remain.

Flat except over the achene: mostly thin and flat except where the achene forces the perigynium to swell; lacking any filling except the achene. (Compare *planoconvex*)

Foliage: leaves, considered collectively.

Fusiform: widest in the middle and tapered to both ends.

Glabrous: not hairy; having a smooth surface.

Glaucous: bluish or grayish in color, due to a waxy layer that can be rubbed off.

Graminoid: grass or grass-like plant (e.g., sedge or rush).

Gynecandrous: with female flowers above male flowers in the same spike. (Compare *androgynous*)

Head: a dense inflorescence of more than one spike, in which individual spikes may be hard to distinguish.

Hyaline: thin, veinless, often ± transparent or whitish.

Impressed: pressed down into, sunken, not raised above.

Indehiscent: incapable of opening (e.g., anthers of most hybrid *Carex*).

Inflated: expanded so that there is space between (e.g., between the perigynium wall and the achene, or between a bract sheath and the culm).

Inflorescence: the entire flowering area at the top of the stem, beginning at the bottom of the sheath of the lowest inflorescence bract.

Inflorescence bract: a bract subtending a spike.

Inframarginal: somewhat in from the margin; an inframarginal rib "ought to" be at the margin of the perigynium but is located away from the edge.

Internode: the part of the culm between nodes, between attachment points of leaves, branches, or spikes.

Involute: with edges rolled upward along the length of the leaf. (Compare *revolute*)

Ladder-fibrillose: (of leaf sheaths) disintegrating and leaving the veins intact in a feather-like pattern with a vertical central fiber and side fibers extending laterally.

Lance-triangular: broadest near the base and tapering gradually to the beak.

Lanceolate: narrow and elongate, widest below the middle. (Compare ovate)

Lateral: at the sides; the lateral spikes are those arising on the side of the culm, not at the tip. (Compare *terminal*)

Lax: (1) tending to bend over. (2) not dense.

Lenticular: two-sided, lens-shaped in cross section; both the surfaces are convex or one is flat. (Compare *trigonous*)

Ligule: a flap of tissue at the base of the leaf blade, on its inner side, next to the culm. In *Carex,* the ligule is mostly fused to the blade, leaving a small rim.

Marginal rib: the thickened rim formed by a vascular bundle along each edge of a perigynium.

Mesic: moderately moist; not wet, not dry.

Midrib: the rib (vascular bundle) in the middle of a bract, scale, or leaf.

Misapplied: (of a name) used for the wrong species, different from the species it was originally intended to name.

Monospecific: composed of a single species.

Montane: in mountains, but below timberline. (Compare *alpine*)

Morphology: structure or external anatomy; or the study of structure.

Mouth of the leaf sheath: the opening at the top of the leaf sheath where the blade attaches.

Mucronate: with a point so short we can't bear to call it an awn.

Naturalized: non-native, established and reproducing in the wild.

Node: the point on the culm where a leaf or a branch is attached. The node of a leaf is located at the bottom of the sheath, not the blade.

Obovate: oval but wider near the top than the base. (Compare *ovate*)

Obtuse: blunt or rounded at the tip, with the sides coming together at an angle wider than 90°.

Orbicular: round and flat.

Ovate: oval and wider at the bottom than the top. (Compare *elliptic, lanceolate, obovate*)

Papillose or papillate: covered with tiny bumps (papillae).

Peduncle: the stalk supporting a spike.

Pendent: (of a spike) dangling, hanging down, because the peduncle bends.

Perigynium (plural, *perigynia*): a specialized bract that surrounds the achene, characteristic of *Carex* and *Kobresia*.

Persistent: staying on the plant, not falling off.

Persistent style: when young, a persistent style has no abrupt change of color. It remains attached to the achene, often becoming bent as the achene grows. (Compare *deciduous style*)

Phyllopodic: with lower leaves of the fertile culm having normal blades; the previous year's dead leaf blades present at the base. (Compare *aphyllopodic*)

Pistillate: female, with an ovary, related to female flowers. A pistillate scale subtends the perigynium. A pistillate spike contains only perigynia, no staminate flowers.

Pithy tissue: soft, air-filled, corky tissue sometimes found at the base of the perigynium or around the achene. It is easy to stick a pin into pithy tissue.

Planoconvex: flat on one side and convex (rounded) on the other. (Compare *flat except over the achene*)

Plant bases: rhizomes and lower parts of the shoots, including any bladeless leaf sheaths.

Plumose: feather-shaped; with a central axis and narrow side branches.

Puberulent: minutely pubescent with very short, fine hairs.

Pubescent: covered with short hairs.

Purplish: in sedges, a dark blackish color with purple or maroon tones.

Rachilla: a vestigial branch found inside the perigynium next to the achene in some *Carex* species.

Recurved: curved downward or outward, curved toward the dorsal surface.

Reflexed: angled backward; pointing more than 90° back from the axis of the spike.

Remote: separated, distant.

Revolute: with edges rolled downward along the length of the leaf. (Compare *involute*)

Rhizomatous: having rhizomes that clearly separate some or all of the culms. (Compare *cespitose*)

Rhizome: an underground stem, usually growing horizontally, often rooting at the nodes, generally with reduced leaves (scales) at the nodes.

Rib: a strong vascular bundle with fibers; usually refers to the two raised, reinforced edges of the perigynium.

Scabrous: rough.

Scale: a small bract subtending a perigynium (pistillate scale) or male flower (staminate scale); or, a much-reduced leaf on a rhizome.

Sedgehead: a person with an inordinate fondness for sedges.

Serrulate: with very small teeth on the edge.

Sessile: lacking a stalk; attached directly on the culm.

Sheath: the lower part of a leaf, which forms a tube around the stem. (Compare *blade*)

Shoulder: as the perigynium curves to become the beak, first its margin curves inward, then outward. If there is a relatively abrupt curve inward (not just a general narrowing), the perigynium has shoulders (e.g., *C. interior*).

Spike: the basic inflorescence unit in *Carex;* an unbranched section of stem with attached scales and perigynia and/or staminate flowers.

Spreading: diverging broadly from the main axis, about 45° to 90°.

Staminate: male; with anthers. The staminate scale is the scale subtending anthers. A staminate spike contains only anthers, no perigynia.

Stigma: the distal surface of the pistil, which sticks out of the perigynium to receive pollen. A *Carex* flower has 2, 3, or rarely 4 stigmas.

Stipe: a small stalk found at the base of an achene or perigynium.

Stipitate: having a stipe.

Stolon: an aboveground stem, rooting at the nodes and bearing shoots.

Style: the stalk-like portion of a pistil, connecting ovary and stigma.

Subalpine: near timberline, but above continuous forest; often in the meadows around the higher clumps of trees,

Subtending: located directly below a plant part and usually more or less covering it, at least when the part is young.

Succulent: fleshy, thickened. (Less apparent in dried plants, because a succulent part withers and becomes hard.)

Suture: seam where two edges are sealed together. Think of the *Carex* perigynium as a once-flat bract that has been curved to wrap around the achene and then sealed. The dorsal suture is the seam where the bract edges sealed together. In some species it is visible, at least on the beak.

Sympatric: (of two or more species) living in the same area.

Taxon (plural, taxa): used to indicate some taxonomic unit without specifying what that unit is. A taxon may be a subspecies, a species, a genus, a family, etc.

Teeth: two projections at the tip of the perigynium beak.

Terminal: uppermost. The terminal spike is the one at the tip of the culm.

Trigonous: (of achenes or perigynia) 3-sided in cross section.

Vein: vascular bundle. On a perigynium, veins are located on the faces and ribs are located at the margins. (Compare *rib*)

Ventral: the surface that is toward the stem, the front surface. (Compare *dorsal*)

Wing: an expanded, flattened edge of a perigynium or culm.

Illustration Credits

We are grateful to the individual artists and photographers who provided illustrations for this book, and to publishers who gave permission to use additional illustrations. It is our intent to credit all photographs and photographers, and we apologize for any omissions or inaccuracies. Illustrations are photographs unless noted otherwise. If all the illustrations for a species account are by the same artist, only the species is listed; otherwise, both the species and the part illustrated are listed. All photographs not credited here were taken by *Carex* Working Group members.

Henry Charles Creutzburg. Drawings from Mackenzie (1940), used with permssion from the New York Botanical Garden. *Carex angustata, C. barbarae, C. bebbii, C. chordorrhiza, C. circinata, C. duriuscula, C. engelmannii, C. feta* (habit), *C. gynodynama, C. integra* (habit), *C. nervina, C. scirpoidea* ssp. *stenochlaena.*

David Dister. *Carex deweyana* fresh perigynia.

Susan Farrington. *Carex eburnea* (habit).

Charles Feddema. Drawings from Hermann (1970). *Carex subfusca.*

John Hilty. *Carex lacustris* inflorescence.

Jeanne R. Janish. Drawings from Hitchcock et al. (1969), used with permission from the University of Washington Press. Species accounts: *Carex albonigra, C. aperta, C. aquatilis* var. *aquatilis, C. aquatilis* var. *dives, C. arcta, C. athrostachya, C. atrosquama, C. brevior, C. buxbaumii, C. capillaris, C. capitata, C. concinna, C. crawei, C. cusickii, C. diandra, C. feta* (inflorescences), *C. fracta, C. integra* (inflorescence, perigynia), *C. interior, C. jonesii, C. leporinella, C. leptalea, C. livida, C. media, C. nardina, C. obtusata, C. paysonis, C. pluriflora, C. proposita, C. retrorsa, C. scirpoidea* ssp. *scirpoidea, C. siccata, C. viridula* ssp. *viridula, Kobresia simpliciuscula.* Introductory text: *Carex atherodes, C. aurea, C. brevior, C. nudata, C. obtusata, C. pauciflora, C. pluriflora, C. proposita, C. raynoldsii, C. rossii, C. rostrata, C. stipata, C. straminiformis, C. subnigricans, C. vulpinoidea, Kobresia myosuroides.*

Emerenciana G. Hurd. Photos from Hurd et al. (1998). *Carex aquatilis* var. *aquatilis* (perigynia, scales, achenes), *C. atrosquama* (inflorescence), *C. capitata* (inflorescence), *C. concinnoides* (inflorescence), *C. crawei* (perigynia), *C. duriuscula* (inflorescence), *C. paysonis* (inflorescence), *C. rossii* (perigynia, scales, and achenes), *C. saxatilis* (inflorescence), *C. vesicaria* (perigynia, scales, achenes).

Richard Heliwell. *Carex heteroneura* (habitat).

Andrew Hipp. *Carex siccata* (inflorescence, inflorescence, habit).

Matthias Hoffmann. *Carex hirta* (2 spikes).

Ernst Horak. *Carex distans* (inflorescence on gray card).

Intermountain Herbarium, Utah State University. *Carex proposita* (habit; specimen).

Emily Kapler. *Carex crawei* (inflorescence).

Tom Kaye. *Carex scabriuscula* (inflorescences).

Natalie Kirchner. *Carex eburnea* (infloresence, perigynium).

Steve Matson. *Carex abrupta* (habit), *C. davyi* (habit), *C. gynodynama* (spike/leaf), *C. harfordii* (perigynium), *C. proposita* (perigynia).

Keir Morse. *Carex mendocinensis* (perigynia and inflorescence).

Anton A. Reznicek. *Carex concinna* (inflorescence), *C. hirsutella* (habit), *C. nigra, C. scoparia* (inflorescences).

Joe Rocchio. *Carex lacustris* (except inflorescence).

Russ Schipper. *Carex deweyana* (habit, inflorescence), *C. scoparia* (inflorescence).

Rena Schlachter. Drawing of *Carex klamathensis.*

Forest and Kim Starr. *Carex longii* (inflorescence and stems).

Amadej Trnkoczy. *Carex distans* (pistillate spike, plant bases).

Dana Visalli. *Carex tenera* var. *tenera* (inflorescence).

Fred Weinmann. *Carex nardina* (inflorescence, habit/habitat, including self-portrait).

Louise Wootton. *Carex kobomugi.*

Gene Yates. *Carex nigricans* (inflorescence).

Peter Zika. *Carex distans* (pair of inflorescences, habit), *C. harfordii* (all but perigynium), *C. hirsutella* (except habit), *C. hirta* (1 spike, foliage, rhizome), *C. subbracteata* (all except perigynia).

References

Only a few of the many references used in preparation of this book are listed here. For a more complete list of references we used, visit our web site, http://www. carexworkinggroup.com.

Arnett, J., and C. Bjork. 1996. Species conservation strategy for *Carex vallicola* Dewey in Washington. Washington Native Plant Society and Okanogan National Forest.

Ball, P. W., and A. A. Reznicek. 2002. *Carex*. pp. 254–572 *in* Flora of North America Editorial Committee, *eds., Flora of North America North of Mexico*, Volume 23: Magnoliophyta: Commelinidae (in part): Cyperaceae. Oxford University Press, New York.

Bell, K. L., and L. C. Bliss. 1979. Autecology of *Kobresia bellardii:* Why winter snow accumulation limits local distribution. *Ecological Monographs* 49: 377–402.

Cronquist, A., A. H. Holmgren, N. H. Holmgren, J. L. Reveal, and P. K. Holmgren. 1977. *Intermountain Flora: Vascular Plants of the Intermountain West, U.S.A.* Vol. 6: Monocotyledons. Columbia University Press, New York.

Decker, K., D. R. Culver, and D. G. Anderson. 2006. *Kobresia simpliciuscula* (Wahlenberg) Mackenzie (simple bog sedge): A Technical Conservation Assessment. Unpublished Report prepared for the USDA Forest Service, Rocky Mountain Region, Species Conservation Project.

Dibble, Alison. 2001. New England Plant Conservation Program Conservation and Research Plan: *Carex atherodes* Sprengel, Awned Sedge. New England Wild Flower Society, 180 Hemenway Road, Framingham, MA 01701. 15 pp. http://www.newfs.org/pdf/Carexatherodes.pdf.

Dunlop, Debra A. 2004. *Carex crawei* Dewey (Crawe's Sedge) Conservation and Research Plan for New England. New England Wild Flower Society, 180 Hemenway Road, Framingham, MA 01701.

Gage, E., and D. J. Cooper. 2006. *Carex diandra* Schrank (lesser panicled sedge): A Technical Conservation Assessment. [Online.] USDA Forest Service, Rocky Mountain Region. Available: http://www.fs.fed.us/r2/projects/scp/assessments/carexdiandra.pdf.

Harris, J. G., and M. W. Harris. 1994. Plant Identification Terminology: An Illustrated Glossary. Spring Lake Publishers, Spring Lake, UT.

Hermann, F. J. 1970. Manual of the Carices of the Rocky Mountains and Colorado Basin. USDA Forest Service Agriculture Handbook No. 374.

Hipp, Andrew L., Anton A. Reznicek, Paul E. Rothrock, and Jaime A. Weber. 2006. Phylogeny and classification of *Carex* section *Ovales* (Cyperaceae). *International Journal of Plant Sciences* 167: 1029–48.

Hitchcock, C. L., A. Cronquist, M. Ownbey, and J.W. Thompson. 1969. *Vascular Plants of the Pacific Northwest*, Part 1: Vascular Cryptogams, Gymnosperms, and Monocotyledons. University of Washington Press, Seattle.

Hitchcock, C. L., and A. Cronquist. 1973. *Flora of the Pacific Northwest*. University of Washington Press, Seattle.

Hurd, E. G., N. L. Shaw, J. Mastrogiuseppe, L. C. Smithman, and S. Goodrich. 1998. Field Guide to Intermountain Sedges. General Technical Report RMRS-GTR-10. USDA Forest Service, Rocky Mountain Research Station, Ogden, UT. 282 pp.

Hutton, E. E. 1976. Dissemination of perigynia in *Carex pauciflora*. *Castanea* 41: 346-48.

Jones, T. M. 2007. *Carex* Interactive Identification Key. http://utc.usu.edu/keys/Carex/Carex.html.

Lesica, P., 1998, Conservation status of *Carex parryana* ssp. *idahoa* in Montana, unpublished report on file at the Bureau of Land Management–Dillon Field Office, Dillon, MT.

Mackenzie, K. K. 1940. *North American Cariceae*. 2 volumes. New York Botanic Garden. Illustrations posted at http://www.csdl.tamu.edu/FLORA/carex/carexout.htm.

Mastrogiuseppe, J. 1993. *Carex*. pp. 1107–38 *in* J. C. Hickman, *ed.*, *The Jepson Manual: Higher Plants of California*. University of California Press, Berkeley.

McClintock, Katherine A., and Marcia J. Waterway. 1994. Genetic differentiation between *Carex lasiocarpa* and *C. pellita* (Cyperaceae) in North America. *American Journal of Botany* 81: 224-31.

Naczi, R. F. C. 2002. Seven new species and one new combination in *Carex* (Cyperaceae) from North America. *Novon* 12: 508–32.

Noble, J. C. 1982. Biological flora of the British Isles: *Carex arenaria* L. *Journal of Ecology* 70: 867-86.

Reznicek, A. A., and P. M. Catling. 1986. Vegetative shoots in the taxonomy of sedges (*Carex*, Cyperaceae). *Taxon* 35: 495-501.

Schütz, W. 2000. Ecology of seed dormancy and germination in sedges (*Carex*). *Perspectives in Plant Ecology, Evolution, and Systematics* 3: 67-89.

Shipley, B., and M. Parent. 1991. Germination responses of 64 wetland species in relation to seed size, minimum time to reproduction and seedling relative growth rate. *Functional Ecology* 5: 111–18.

Standley, L. A. 1985. Systematics of the *Acutae* group of *Carex* (Cyperaceae) in the Pacific Northwest. *Systematic Botany Monographs* 7: 1-106.

Stevens, M. L. 2004. Whiteroot (*Carex barbarae*). *Fremontia* 32: 3–6.

Stotts, B. 1992. Monitoring Plan for Henderson's Sedge (*Carex hendersonii*) in the Big Smith Analysis Area. Lochsa Ranger District, Clearwater National Forest, Kooskia, ID.

Wilson, B. L., R. Brainerd, M. Huso, K. Kuykendall, D. Lytjen, B. Newhouse, N. Otting, S. Sundberg, and P. Zika. 1999. Atlas of Oregon *Carex*. Native Plant Society of Oregon, Occasional Paper No. 1, Corvallis, Oregon. 29 pp.

Zika, P. F., A. L. Hipp, and J. Masrogiuseppe. 2013. *Carex*. pp. 1308-38 in B. G. Baldwin, D. H. Goldman, D. J. Keil, R. Patterson, T. J. Rosatti, and Dieter H. Wilken, eds., *The Jepson Manual: Higher Plants of California*, 2nd Edition. University of California Press, Berkeley.

Index

Species accounts are in bold type